Planetary Gentrification

Planetary Gentrification

Loretta Lees, Hyun Bang Shin and
Ernesto López-Morales

polity

Copyright © Loretta Lees, Hyun Bang Shin and Ernesto López-Morales 2016

The right of Loretta Lees, Hyun Bang Shin and Ernesto López-Morales to be identified as Authors of this Work has been asserted in accordance with the UK Copyright, Designs and Patents Act 1988.

First published in 2016 by Polity Press

Polity Press
65 Bridge Street
Cambridge CB2 1UR, UK

Polity Press
350 Main Street
Malden, MA 02148, USA

All rights reserved. Except for the quotation of short passages for the purpose of criticism and review, no part of this publication may be reproduced, stored in a retrieval system, or transmitted, in any form or by any means, electronic, mechanical, photocopying, recording or otherwise, without the prior permission of the publisher.

ISBN-13: 978-0-7456-7164-2 (hardback)
ISBN-13: 978-0-7456-7165-9 (paperback)

A catalogue record for this book is available from the British Library.

Library of Congress Cataloging-in-Publication Data
Lees, Loretta, author
    Planetary gentrification / Loretta Lees, Hyun Bang Shin, Ernesto López-Morales.
        pages cm. – (Urban futures)
    Includes bibliographical references and index.
    ISBN 978-0-7456-7164-2 (hardcover : alk. paper) – ISBN 0-7456-7164-0 (hardcover : alk. paper) – ISBN 978-0-7456-7165-9 (pbk. : alk. paper) – ISBN 0-7456-7165-9 (pbk. : alk. paper) 1. Gentrification. 2. Urbanization. 3. Urban geography. I. Shin, Hyun Bang, author. II. López Morales, Ernesto, author. III. Title.
    HT170.L443 2016
    307.2–dc23
                                                                                2015027160

Typeset in 11.5 on 15 pt Adobe Jenson Pro by Toppan Best-set Premedia Limited
Printed and bound in the UK by CPI Group (UK) Ltd, Croydon

The publisher has used its best endeavours to ensure that the URLs for external websites referred to in this book are correct and active at the time of going to press. However, the publisher has no responsibility for the websites and can make no guarantee that a site will remain live or that the content is or will remain appropriate.

Every effort has been made to trace all copyright holders, but if any have been inadvertently overlooked the publisher will be pleased to include any necessary credits in any subsequent reprint or edition.

For further information on Polity, visit our website: politybooks.com

Contents

Acknowledgements	vii
1. Introduction	1
2. New Urbanizations	24
3. New Economics	53
4. Global Gentrifiers: Class, Capital, State	83
5. A Global Gentrification Blueprint?	111
6. Slum Gentrification	140
7. Mega-Gentrification and Displacement	171
8. Conclusion	201
References	227
Index	256

Acknowledgements

Writing this book has been an illuminating experience, as we went through a wide range of both early and recent writings on gentrification and urbanization around the world. The project received a boost from a series of workshops on global gentrifications, which took place in London and Santiago de Chile in March and April 2012 respectively. We would like to thank all the participants in those workshops, and especially Hilda Herzer, who initially participated in this comparative project before she sadly passed away in 2012. The funding from the Urban Studies Foundation and the Urban Studies journal for these events was greatly appreciated. We also thank all those who have kindly provided images other than our own.

Our individual acknowledgements are as follows:
Loretta Lees. I would like to thank David Ley for discussions on global gentrifications on his recent visit to the UK; Jenny Robinson for inspiring my interest in comparative urbanism; Sandra Annunziata for working with me on a Marie Curie project on resistance to gentrification in Southern European cities; my past and current PhD students working on gentrification around the globe. Thanks also to Tom Slater and Elvin Wyly for their continued support. Special thanks to my partner David and daughters Meg and Alice, who had to do without me while I worked on the text for the book; and to Hyun and Ernesto who have been wonderful collaborators and comrades.

Hyun Bang Shin. I would like to thank all the colleagues whose kind invitations to speak on various occasions provided key inspiration for the writing of this book. Particular thanks are to Soo-hyun Kim and Bae-Gyoon Park for their continued interest and support for this work, and to my former and current students for lending their critical ears. The National Research Foundation of Korea Grant funded by the Korean Government (NRF-2014S1A3A2044551) is also acknowledged for its support for part of the works discussed in this book. Special thanks are to Loretta and Ernesto whose solidarity and dedication provided the much needed push for this book. Most of all, I express my deepest love and gratitude to my partner Soo-Jeong for her love and patience while the book was being drafted, and for providing inspiration as always for my research.

Ernesto López-Morales. I would like to thank both the COES – Centre for Social Conflict and Cohesion Studies (CONICYT/FONDAP/15130009) and the Contested Cities Project (People Marie Curie Actions Projects, Code: PIRSES-GA-2012-318944), for helping to fund several of the case studies presented in this book. Special thanks to my friends, colleagues and students from the University of Chile who support and intelligently criticize my work. Also appreciation to my encouraging partner Ignacia, my son Martín, and especially co-authors Loretta and Hyun for bringing this collaborative project to its highest level.

1 Introduction

As readers of the gentrification literature will know, the British sociologist Ruth Glass coined the term 'gentrification' in 1964 in her book 'London: Aspects of Change':

> One by one, many of the working class quarters of London have been invaded by the middle classes – upper and lower. Shabby, modest mews and cottages – two rooms up and two down – have been taken over, when their leases have expired, and have become elegant, expensive residences. Larger Victorian houses, downgraded in an earlier or recent period – which were used as lodging houses or were otherwise in multiple occupation – have been upgraded once again. Nowadays, many of these houses are being subdivided into costly flats or 'houselets' (in terms of the new real estate snob jargon). The current social status and value of such dwellings are frequently in inverse relation to their status, and in any case enormously inflated by comparison with previous levels in their neighbourhoods. Once this process of *'gentrification'* starts in a district it goes on rapidly until all or most of the original working class occupiers are displaced and the social character of the district is changed. (Glass 1964a: xviii–xix, italics added).

Ruth Glass's other writings, however, including those on urbanization outside of Britain are much less well known. For example, that same year she wrote about the 'Gaps in Knowledge' in studies of urbanization in non-Western contexts:

> So far, our knowledge of the current processes, configurations and implications of urbanization in the developing countries has been limited, or even apparently arrested, in several interrelated respects. First, the framework of analysis and enquiry in this field (as in many others) has been heavily conditioned by Western, and particularly Anglo-Saxon, experience – or rather by categories of thought derived from the as yet inadequately documented, only sketchily compared and partially interpreted, history of nineteenth and early twentieth urbanization in the now industrialized countries, notably Britain and the United States. It is partly because the 'shock of urbanization' felt in these countries during earlier periods is still reverberating, that the notions formed under its impact, whether expressed in terms of reason or unreason, have remained so tenacious and pervasive. The influence of such notions is reflected in the choice of subjects with which students of contemporary urban growth and phenomena in the developing countries have been preoccupied. The predominance of Western thought, in general, is reflected in the treatment of such subjects, which tends to follow both the conventional lines of demarcation between matters urban and rural, and also the established boundaries between the various disciplines of the social sciences. (Glass 1964b: 1–2)

Hers was a prescient 'comparative urbanism' that was concerned about the dominance of Western thought and experience in studies of urbanization in developing countries, what Ma and Wu (2005: 10–12) have called a Western-centric 'convergence thesis'. Other Marxists, for example Henri Lefebvre (2003: 29), had similar concerns about the hegemony of the Euro-American *industrialized* city in urban theory:

> We focus attentively on the new field, the urban, but we see it with eyes, with concepts, that were shaped by the practices and theories of

industrialization, with a fragmentary analytical tool that was designed during the industrial period and is therefore reductive of the emerging reality.

In this book, we take on board a 'new' comparative urbanism (see Robinson 2006, 2011a) to address the concern that over the past two decades the term 'gentrification' itself has been conceptually stretched to uncritically assume a similar trajectory around the globe (see Lees 2012). It is 'new' because it focuses on cities beyond the usual suspects of London, New York, etc., and beyond the constructs that have come out of, or been based on, those places. As Ley and Teo (2014: 1286) argue, this conceptual overreach 'represents another example of Anglo-American hegemony asserting the primacy of its concepts in other societies and cultures'.

This book is one of the first to unpack this hegemony and to question the notion of a 'global gentrification'. Glass (1964b: 18) goes on to ask: 'What happens to the elaborate theories and speculations on the trends and implications of urbanization on the international scale when it has to be admitted that even the most elementary raw material for their verification exists?'

In our unpacking of the notion of a 'global gentrification', we discuss gentrification beyond the usual suspects in Britain, Europe and North America, gathering in raw material on processes that have been labelled 'gentrification' in non-Western cities and on processes that have not been labelled as 'gentrification'. In so doing, we consider the extent to which Western theorizing on gentrification can be useful in non-Western cities. For, like Glass (1964b: 27) we are conscious of the 'persistence of the Western ideology of urbanism (or rather anti-urbanism)' which may not exist (or at least not in the same way) in non-Western contexts where, for example, issues of informality, state developmentalism (often intertwined with advanced neoliberalism), and even the concept of neighbourhood itself, take on radically different meanings.

Building upon recent urban studies scholarship that has revisited the concept of the urban and the process of urbanization at multiple scales (see Merrifield 2013a; Brenner and Schmid 2012, 2014; Keil 2013), we advance the view that gentrification is becoming increasingly influential and unfolds at a planetary scale. This foray into 'planetary gentrification' advances postcolonial geographies along some of the pathways that Sidaway et al. (2014) suggest, for in this book we: (i) narrate planetary gentrifications and the configurations *between* their paths, focusing on the ascendancy of the secondary circuit of real estate (Harvey 1978; Lefebvre 2003) (we offer a global perspective that considers colonialisms, analytical and everyday comparativisms, globalization, and also the globalized effects of financial capitalism), (ii) we acknowledge other (post)colonialisms (old and new), (iii) we demonstrate planetary indigeneity (organic gentrifications that are not copies of those in the West), and (iv) we problematize translations (West to East, North to South and vice versa).

Gentrification is argued to have 'gone global', to have spread geographically – what the late Neil Smith (2002) called 'gentrification generalized'. Atkinson and Bridge (2005: 1; italics added) have proclaimed that 'Gentrification *is now* global' and gone on to discuss gentrification as the new urban colonialism in a global context. In arguing that gentrification has gone global, they assume a North to South, West to East trajectory, and that gentrification has moved down the urban hierarchy from First World to Second and Third World cities. They also assume that gentrification is not indigenous to these contexts and that it is new to them. The global is seen as originating from the West. Blaut (1993: 12) argues that such diffusionist thinking is an example of 'spatial elitism' that inscribes a geography of centre and periphery on the world. By way of contrast, others, for example Tim Butler (2007a, 2010), are concerned that lots of different changes are becoming subsumed under the 'gentrification brand' and as such the concept has become 'diluted' and we are 'losing sight of what it is that needs to be explained or at least understood'. Indeed, Sharon Zukin (2010: 9) argues that 'gentrification generalized' is really a broad process of 're-urbanization' in which city space is taken up by white collar men and women and their consumption tastes and habits, creating an economic division but also a cultural barrier between rich and poor, young and old; her research focus though is in the West – New York City – again!

By way of contrast, this book begins the process of ontological awakening to the process of gentrification in cities outside of the Euro-American heartland, in so doing we consider the claim that gentrification 'has gone' global, the idea that gentrification is a 'force' that has travelled or diffused outwards from a certain 'centre' towards global peripheries. We show that gentrification is a phenomenon that cities worldwide have experienced (it is not totally new in the twenty-first century to the global South) and are experiencing (through different types of urban restructuring).

There are material issues at the moment of co-writing a book like this. We draw on: (i) our regional and linguistic expertise (Lees on Europe and North America [languages English, and some German and French], Shin on Southeast and East Asia [languages Korean, Chinese and English], and López-Morales on Latin America [languages Spanish, English, Portuguese); (ii) the information we collected in the workshops we ran on global gentrification two years ago, that went into producing an edited collection on global gentrifications (see Lees, Shin and López-Morales 2015) and two regionally focused special issues – one on East Asia (*Urban Studies* 2016) and one on Latin America (*Urban Geography* forthcoming); and (iii) a survey of the various non-Western case studies of gentrification (in countries as diverse as China, South Korea, India, Brazil, Chile and South Africa) that have begun to emerge in the twenty-first century. We have done the learning that McFarlane (2011) asks for, one that actively involves bringing together assemblages of 'people-sources-knowledges' to expose and unlearn existing conceptualizations/theories, ideologies and practices/policies.

We have done the comparative urbanism or transurban learning that underpins this book 'together', but it helped that we all share/d the same approach – critical political economy. We are concerned with uneven spatial development in cities and the modes of regulation that manage capitalism in cities, especially in its current phase. The domination of capitalist interests continues to shut down alternatives to gentrification. Although the theory behind critical political economy has been produced in the context of Euro-American cities, as Roy (2009: 825) argues, 'this is not to say that this analysis is not applicable to the cities of the global South. Indeed, it is highly relevant.' In fact, it would be naïve to claim North–South cultural and theoretical exchanges are a recent problem. Capitalism has unfolded in the South following its own trajectory of development, and major contributions from Marxism and liberalism in the South have input into theories of state developmentalism, dependency, and marginality (three useful concepts that still help to analyse urban change in many places). What we have to be alert to though are the different ways in which the uneven production of urban space, the production of differentiated spatial value, takes place in non-Western cities. Ours then is an open, embedded and relational understanding of gentrification, a stance that is (as we say) historical, and that draws on Massey (1993: 64):

> interdependence [of all places] and uniqueness [of individual places] can be understood as two sides of the same coin, in which two fundamental geographical concepts – uneven development and the identity of place – can be held in tension with each other and can each contribute to the explanation of the other.

On the other hand, in unpacking 'global gentrification' we also draw on a recent wave of scholarship on postcolonial urbanism that seeks to unhinge, unsettle, contextualize or 'provincialize' Western notions of urban development. Like Glass (1964b) we see the need to breach the divide between what was until recently called 'development studies' and urban studies, which has long been dominated by Western scholarship. This means unpacking Western-based approaches, including being more careful in dealing with theories on neoliberal urbanism. More recently, Jennifer Robinson (2002) has identified similar issues to Ruth Glass, identifying a geographical division between urban studies and theory focused on the West and development studies focused on what were once known as 'Third World cities'. The result of the overlapping dualisms 'theory'/West and 'development'/Third World is that 'urban studies is deeply divided against itself' (Robinson, 2002: 533), and this narrows the vitality and purchase of urban theory and has consequences for urban policy. Robinson (2003) calls this 'asymmetrical ignorance', and this book seeks to overcome that ignorance in looking at a particular process – gentrification – globally. This should not be misread as meaning that scholars in the global South are ignorant about the process they study, or incapable of producing theorizations about those processes. On the contrary, in analysing the currently available literature on gentrification 'beyond the usual suspects', we also focus very much on the hypotheses that have been constructed by local authors.

Gentrification studies has long been at the forefront of opening up and moving beyond the traditional dichotomies of urban studies – from its rejection of the ecological urban models of the Chicago School of Sociology to discussions of rural and suburban gentrification which have already demonstrated the extension of 'the urban world' beyond the city and the inner city (see Chapter 4 in this volume). As such, gentrification researchers are well positioned not just to dispense with the old binaries of city and suburb, urban and rural, but also between North and South, developed and developing worlds (Lees 2012, 2014a). To some extent, our book is a response to Andy

Merrifield's (2014) recent call for 'a reloaded urban studies', which calls for the removal of centre-periphery binary thinking, acknowledging the emergence of multiple centralities across urbanizing spaces and 'dispens[ing] with all the old chestnuts between global North and global South, between developed and underdeveloped worlds, between urban and rural, between urban and regional, between city and suburb, just as we need to dispense with old distinctions between public and private, state and economy, and politics and technocracy' (Merrifield 2014: 4).

Although the postcolonial urban critique that we undertake in this book means 'unlearning' what we have learnt (Spivak 1993 in Lees 2012) it also necessitates, we would argue, not throwing away what we have already learned from more established (Western) urban theories in gentrification studies. Instead, we ask which elements in the South as well as the North could enrich gentrification theory and concepts. We follow Ananya Roy (2009: 820; see also Parnell 1997, who made a similar point) in this regard:

> The critique of the EuroAmerican hegemony of urban theory is thus not an argument about the inapplicability of the EuroAmerican ideas to the cities of the global South. It is not worthwhile to police the borders across which ideas, policies, and practices flow and mutate. The concern is with the limited sites at which theoretical production is currently theorized and with the failure of imagination and epistemology that is thus engendered.

Despite our interest in Glass's writings, we do not, however, follow Glass's definition of gentrification. For as Lees, Slater and Wyly (2008) show, the process of gentrification has mutated so much over time to make that definition rather dated. As Beauregard (2003: 190) has said by holding one city up as a model, in this case gentrification as Glass's London in the 1960s, 'comparative analysis is reduced to a perfunctory and unenlightening assessment of how the "others" compare'.

Instead, we follow Clark's (2005: 258) more recent and expansive definition that is not tied to the experience of a particular city at a particular time:

> Gentrification is a process involving a change in the population of land-users such that the new users are of a higher socio-economic status than the previous users, together with an associated change in the built environment through a reinvestment in fixed capital. The greater the difference in socio-economic status, the more noticeable the process, not least because the more powerful the new users are, the more marked will be the concomitant change in the built environment. It does not matter where, it does not matter when. Any process of change fitting this description is, to my understanding, gentrification.

As Robinson (2011a: 17) reminds us, 'the most abstract concepts offer an opportunity to incorporate the widest range of cities within comparative reflection. Abstract concepts are also the level at which urban theory is most open to a creative generation of concepts that might help us look differently at cities and their problems ... urban studies could find in the empirical, comparative interrogation of its most abstract concepts a rich field for creative reconceptualization.'

Gentrification for us, like for Zukin (2010), is a displacement process, where wealthier people displace poorer people, and diversity is replaced by social and cultural homogeneity. This we believe undermines urbanity and the future of cities as emancipatory places. As Betancur (2014: 3) points out, some authors have built their careers by denying displacement – for example, Freeman (2006), Hamnett (2003) and Vigdor (2002) – despite the obvious 'class substitution' involved. Notably, these are all studies from the 'global North'. Displacement goes beyond 'physical' displacement of residents from their dwellings, and encompasses the phenomenological displacement (see Davidson and Lees 2010) that occurs due to the increase in displacement pressures as neighbourhoods change their characteristics and the way of life of the previous inhabitants faces extinction. Moreover, it is not socially just. Gentrification in the West, in cities like London and New York, now limits the possibilities of urbanity (see Lees 2004 on these possibilities). Ironically, what seems to be happening in the West now is a kind of suburban-like gentrification where the vitality of the city has become hybridized with the comforts of suburbanization, creating a kind of third space of 'sub-urbanity' in which 'bourgeois bohemians' live. This blurring of urban and suburban in contemporary gentrification processes in cities like London and New York reminds us of the way that early gentrification. Indeed the term itself was associated with the rural. As Lees, Slater and Wyly (2008) point out, Glass's coining of the term 'gentrification' was ironic in that it made fun of the snobbish pretensions of affluent middle-class households who really wanted a rural, traditional way of life but did not have the chance to do so. In similar vein contemporary gentrifiers in Western cities want the excitement and diversity of urbanity but a sanitized, suburbanized, version of it. The big question is: do Western ideologies of urban, suburban and rural have any purchase in non-Western contexts? For, as Roy (2009) argues, epistemes embedded in a singular model of industrialization related to modernity and development are outmoded.

In Glass's (1964b) 'Gaps in Knowledge', she argued: 'apparently, in most areas of the world urban-rural differences are becoming more inconsistent, and rather faint – though for varying reasons' (p. 5). This adds another dimension to the equation. Indeed, Glass was already ahead of the conceptual game, in pointing towards what some urban scholars have called 'planetary urbanization' (see Brenner and Schmid 2012; Merrifield 2013a) where the distinctions between urban and rural have broken down as we have all become urban. By 2050, more than three quarters of the world's population is predicted to be urban, this is what Merrifield (2013b), building upon Henri Lefebvre (2003), calls the final frontier – the complete urbanization of society, or what Brenner and Schmid (2012) call the totalization of capital. Some might argue that gentrification going global is an example of planetary urbanization. Yet others, for example, Roger Keil (2013), are now arguing that in a world of cities, suburbanization is the most visible and pervasive phenomenon.

While we agree with Keil that suburbanization is growing around the world on megacity peripheries from Istanbul to Shanghai and it is deserving of study, we would argue, building upon Merrifield (2014), that urbanization around the world is seeing the production of multiple centralities, forcing us to rethink the traditional singular centrality (inner-city, central city or historic core) of urban development, and also the traditional assumption of gentrification as an inner-city process. We argue that processes of planetary gentrification in cities around the world are producing plural sites of contention as capital accumulation and its spatial fix produce concentrated forms of the urban in historic urban, suburban and rural territories. These processes also take place in the context of making and remaking of the urban and the rural, and of their redefined relationships (see Brenner and Schmid 2015; Walker 2015). It is evident in cases like Santiago (López-Morales forthcoming), Seoul (Shin and Kim 2015) or Washington DC (Mueller 2014) that the redevelopment of low-income neighbourhoods is the most salient housing issue there, if measured quantitatively, let alone qualitatively. Our focus then, for the most part, is on urban gentrifications around the globe where contentions have escalated due to an assault from the state and capital, which has endangered settlements and neighbourhoods that serve the urban poor, as well as those factions of the middle classes who are falling into poverty due to economic restructuring. Why the central city continues to be important globally will become apparent as you read this book, but here, centrality does not correspond to a singular centrality as was assumed by the concentric zone model once espoused by the Chicago School. The location of gentrification in planetary urban debates vindicates the enduring interest in gentrification as being at the cutting edge of urban studies and the title of this book: 'Planetary Gentrification'. Importantly, we move beyond the usual gentrification suspects (e.g. London and New York City) to present a picture of urbanization as gentrification. We agree with Smith (2002) that gentrification is the leading edge of global urbanism, at least for now, but it is the leading edge beyond the usual suspects, and this is closely correlated with the ways in which contemporary capitalism raises the status of speculation in real estate in particular, not only in the global North but increasingly in the global South too (e.g. Goldman 2011; Desai and Loftus 2013; Shin 2014a, 2015).

The book is part of an emergent 'cosmopolitan turn' in urban studies which seeks a truly global urban studies. We follow this in seeking a truly global gentrification studies. This necessitates what Heidegger (1927/1996) calls 'de-distancing' and Spivak (1985) calls 'worlding' – the art of being global, of looking at the distinctive experiences of non-Western cities (see also Roy and Ong 2011). As Roy (2009) has said:

> While the twentieth century closed with debate and controversy about the shift from a "Chicago School" of urban sociology to the "Los Angeles School" of postmodern geography, the urban future already lay elsewhere: in the cities of the global South, in cities such as Shanghai, Cairo, Mumbai, Mexico City, Rio de Janeiro, Dakar, and Johannesburg. Can the experiences of these cities reconfigure the theoretical heartland of urban and metropolitan analysis? (p. 820)

Work by 'new' comparative urbanists, like Jennifer Robinson, Ananya Roy, AbdouMaliq Simone, Susan Parnell, Colin McFarlane, and others, is often lumped together even if there are subtle differences between their works. Comparative urbanists like Robinson and Parnell do not want a central urban theory, rather they want theory to be emergent in different places, making it more agile and flexible. They desire different analytical potentials for conversations and want to take on big concepts. Their comparative thinking is about changing theories and understandings: it is about new practices of theorizing, which in turn reshapes our intellectual practices. We do not need to travel, they argue, because in doing so we enact colonialism: rather what is required is a collegial production of knowledge. Such a collegial production of knowledge underwrites this book. McFarlane (2006) posits a 'strategy of critique' that reveals the distinctiveness of urban theories, like gentrification theories, and a 'strategy of alterity' that generates new ideas, lines of enquiry and positions. For McFarlane (2006), like ourselves, comparison is learning across 'the North-South divide'.

The comparative urbanism that we do in this book is not simply the systematic study of similarity and difference among cities in terms of like processes, rather it 'addresses descriptive and explanatory questions about the extent and manner of similarity and difference' (Nijman 2007: 1). We are focusing on gentrification as an urban process, less an urban form, even if they are interrelated. Ours then is not the comparative study that Robinson (2002) rejects (see Table 1.1): ours is a transnational examination that uses one site to pose questions of another (Roy 2003: 466). We perform Robinson's (2011a) 'comparative gesture', but at the same time we try hard to avoid academic impressionism (Lees in press). Ours is a relational comparative approach that acknowledges both the territorial and relational geographies of cities (see Ward, 2009). This involves looking at how cities' pasts, presents and futures are implicated in each other, posing questions of each other. As Hart (2004: 91) has argued, we need to come 'to grips with persistently diverse but increasingly *interconnected* trajectories of sociospatial change in different parts of the world'. Before Lefebvre, Marxism attempted this goal but it failed. Currently, given our critical political economy backgrounds a relational comparative approach to gentrification globally makes a lot of sense, given the increasingly neoliberalized and interconnected world in which we reside. It is an approach, though, that desires to theorize back, reflecting, as we do throughout the book, but especially in the

Table 1.1 Traditional comparisons of gentrification versus a relational comparative approach

Traditional comparisons in gentrification studies
The city is bounded.
The city is a given.
The singularity of cities.
The neighbourhood is bounded.
The neighbourhood scale relates directly to the city scale.
Similarities and differences are used to back up theory and project theory.
Theorizing/conceptualizing from single case studies.
Theory building is certain.
A relational comparative approach in gentrification studies
The city is unbounded.
The city is constituted through its relationships (flows and networks) with other places.
The multiplicity and diversity of cities and their centralities (including renewed centre-periphery relationships).
The neighbourhood is constituted through its relationships with other places.
The neighbourhood, city, regional and global scales are inter-scalar and politicized.
Similarities and differences are used to theorize back and check/change theory.
Theorizing/conceptualizing beyond single case studies.
Theory building is more tentative and evolving.

conclusion, on what this means for existing theories on gentrification. Importantly it can only be an ongoing conversation across cities around the world, a conversation whose ultimate goal is social justice for all.

Like the 'Subaltern Studies School' (e.g. Chakrabarty 2000) which questioned universalizing Western Marxist categories for studying historical social and economic change in South Asia but who wanted to retain a Marxist analysis, we too separate ourselves from Marxism's universalist history of capital, the nation and the political, and from readings of class consciousness that do not travel well to contexts outside of the industrialized West. And like them, we also insist on a Marxist focus on the struggles of subaltern groups, the oppressed and the alienated in urbanizing societies, that aligns with our critical political economy approach. The book also argues that the role of the state has been under-conceptualized in gentrification studies to date and in so doing shows how urban governance in metropolises in the global South has entered what Schindler (2015) calls 'a territorial moment' in which municipal governments are increasingly focusing on transforming urban space rather than improving populations (even if the latter still happens to different degrees in different places and the latter is also used to 'sell' the former as upgrading for the population as a whole). This shift has been driven by political economic circumstances, and Schindler (2015) argues that any attempt now to 'reload' urban studies (see Merrifield, 2014) must focus on 'the governance of territory – i.e. the reconfiguration of power and place – in metropolises at the frontier of the urban revolution' (p. 7). So, what is fostering this shift or moment? Schindler (2015: 14) makes some useful points – that elites, not always the 'middle classes', prefer to invest in real estate in the global South rather than in productive sectors of the economy because there is a disconnect between capital and labour. As he says:

> residents of, say, Lagos, Jakarta or Istanbul, may reasonably assume that in cities of such size they will be able to find a buyer for a luxury apartment in the future, while producing commodities – for domestic consumption or for export – is perceived as risky in comparison. Finally, middle classes in developing countries are not only local beneficiaries of the global regime of open markets and internationalized production,

but according to Ballard (2012, p. 567) they enjoy 'almost entirely positive and unproblematic connotations' among many development agencies and governments. Thus, the construction of infrastructure and the development of a regulatory framework that encourages urban renewal and investment in real estate can be interpreted as attempts to 'reinforce the conditions for their further accumulation'. (Ballard, 2012, p. 569)

In brief, governments in Southern metropolises are excited at the possibility of accumulating capital while remaking their cities.[4] Especially in the context of pursuing industrial production as well as the remaking of cities, the spatial fix as a remedy for over-accumulation crises may not be what is happening. As Shin (2014a: 511–12) states in relation to China's speculative urbanization, 'it is not simply the over-accumulation in the primary circuit of industrial production which facilitates the channeling of fixed asset investment into the secondary circuit of the built environment. Both circuits reinforce each other's advancement, while the state monopoly of financial instruments provides governments and state (and state-affiliated) enterprises the possibility of tapping into the necessary finances.'

Remaking cities in Southern metropolises is done in different ways. The most subtle form of governance has been around *the regulation and securitization of space* – from slum pacification programmes and slum tourism in Rio de Janeiro (see Cummings 2015), to moving *ambulantes* off central city streets in Mexico City (see Lees 2014a); often, this is a sanitization of space in order to attract tourists and it leads to 'touristification' (see Lees, Shin and López-Morales, 2015). As Peck and Theodore (2010a: 172) say, there are deep-seated historical connections between Mexico City, Washington DC and New York City facilitated through the World Bank and Inter-American Development Bank that give policy advice on such revanchist programmes. This can be considered to be another example of Clarke's (2012) 'actually existing comparative urbanism'. *Megaprojects* are another way that governments seek to accumulate capital while remaking their cities. Schindler mentions the Eko Atlantic megaproject in Lagos, Nigeria, which aims to reclaim submerged coastline for a new district – Bill Clinton's inauguration comments sum up the overinflated aspirations of such projects:

> It will work to improve the economy of Nigeria. All over the world, it will bring enormous opportunities. I am convinced that within five years, people will be coming from all over the world to see this [retaining] wall. (Akinsami 2013, cited in Schindler 2015: 16)

Some of these megaprojects are more temporary, for example, those associated with mega events like the Olympics, World Cup or Commonwealth games. Others are more permanent. However, they commonly produce lasting impacts on the social fabric of host cities and beyond, as governments make use of these mega-events as a means to initiate more permanent urban spatial restructuring. Shin (2012, 2014b) discusses how mega events such as the 2008 Beijing Olympics, the 2010 Shanghai World Expo, and the 2010 Guangzhou Asian games became the catalysts to spatial restructuring in respective host cities and further accumulation of fixed assets at the expense of the loss of affordable homes and housing rights of affected residents (see also Davis 2011 for the case of the 1988 Seoul Olympic Games). Such processes provide breeding grounds for gentrification. In cities with global aspirations, there have also been more *comprehensive transformations*, as seen in the global or world city visions of Shanghai, Mumbai or Dubai. All of these different strategies of accumulating capital have meant that capitalism has rendered parts of the population disposable, and led to large and small-scale displacements. Yet, ironically some of these large-scale urban renewal (read gentrification) schemes are supported by the very populations they dispossess (see

Doshi 2015 and Ley and Teo 2014). So, the question of new forms of resistance to these state-led forms of gentrification in the global South becomes quite interesting. The key question is, as Schindler (2015: 21) puts it so well:

> Mainstream Marxian theory narrates how this class [the proletariat] became a class-for-itself in the context of being collectively alienated from means of production. But how do urban residents understand their place in the city, either individually or collectively, if they cannot realistically conceive of selling their labor power for a wage in an era of disconnected capital and labor? Are residents in twenty-first-century metropolises subjectified by regimes of urban transformation in ways that activate them to participate in the transformation of cities? Or does antagonism over access to urban space, infrastructure, and material flows of resources produce a collective consciousness the way that struggles on the shopfloor once did?'

Policymakers worldwide are more interested in their new urban imaginations than they are in providing labour access to jobs. As Harvey (2012) has said, we need a different definition of the proletariat now, of what they are and where they want to go. Together with the reconceptualization of what the urban means, this forces us to think of urban social movements and the question of urban rights in the contemporary world.

Planetary urbanization unfolds in a multi-faceted way, involving a diverse set of agencies and actors that have a stake in accumulation and sustenance of class power. In particular, planetary urbanization plays out in the form of 'accumulation by dispossession'. Harvey's (2007) upgrading of 'primitive accumulation' for the twenty-first-century neoliberal context involves, we argue, state-led gentrification taking place more often than not on formerly public land. Like Merrifield (2013a, 2013b), we prefer the term *planetary* as it suggests something more vivid and growing than the moribund global. (Re)investment in the secondary circuit of capital (the built environment, real estate) is key to this process and incurs a range of dispossessions, of which gentrification constitutes a major part. In some parts of the global South, (re)investment in the secondary circuit is happening at the same time as investment in the primary circuit of capital (industrial production), for example, in China (see Shin 2014a). In other places, it is triumphing (re)investment in the primary circuit (for example, Dubai). Importantly, with planetary urbanization, 'rural places and suburban spaces have become integral moments of neoindustrial production and financial speculation, getting absorbed and reconfigured into new world-regional zones of exploitation, into megalopolitan regional systems . . .' (Merrifield, 2013b: 10). One key issue that Merrifield (2013a, 2013b) and others have paid less attention to, however, is the double character of access to land and housing as both a commodity and a social right (part of the postwar social contract – which is breaking down) in Western capitalist society and how this is not evident in non-Western cities, at least not in the same way. Ley and Teo (2014) discuss this in relation to whether or not gentrification exists in Hong Kong, and how the 'culture of property' influences people's understanding of their housing rights.

Our challenge was to access the different social, economic and political histories of different places and the different languages of gentrification and how they reflect processes of gentrification in different places. But beyond these differences we needed to get to grips with the city as both territorial and relational (with links to elsewhere). We began by thinking, in broad regional terms, about what we knew about gentrification globally:

In *Europe*, despite the hegemony of 'Western' understandings of gentrification, it is important to note that there is not as cohesive a 'Northern' or 'Western' idea about gentrification as one might presuppose. In Paris, for example, whose central city has long been middle class, the result of Haussmannization, sociologists like Préteceille (2007) have argued that the city is experiencing 'embourgeosiement' not gentrification. In addition, gentrification did not emerge post-war (in the 1950s and 1960s) like it did in London in the rest of Europe. For example, in Germany, the term 'gentrification' – *die gentrifikation* or *die gentrifizierung* – has only really emerged in public discourse more recently, at the same time as in the global South!

In the 1990s, Ward (1993) argued that urbanization and inner-city development in *Latin America*, although tied to processes which are global in terms of their economic genesis and consumerist inspiration, have produced rather different outcomes from those observed in North America and in the United Kingdom. He even identified a 'modest gentrification' and population displacement, even if he drew on very few cases. More recently there has been discussion about the relevance of Euro-American gentrification and the idea of a specifically Latin American style of gentrification (see Janoschka, Sequera and Salinas, 2014) that could link Mexico, Brazil, Chile and Argentina, due to the colonial legacy of Spain and Portugal. This work could be considered to be part of a longer tradition in Latin American theory – the work of *dependentistas* (Quijano 1968; Cardoso and Faletto 1979; Biel 2000), who have acted as a counterpoint to modernization and developmentalist theories. In addition, Latin Americans are very concerned with the conditions of urban citizenship, marginality and inequality in access to land and services; and intrigued by the possibilities of 'insurgent citizenship' (Roy 2009; Holston 2008). There is certainly very vocal anti-gentrification activism in Latin America, especially Chile, and the academic side of this has been heavily influenced by Manuel Castells' (1973) idea of the city as a space of social movements and mobilization.

The *East Asian* literature on gentrification to date has been limited in the sense that gentrification per se has not often been invoked in discussions of the urban spatial restructuring and residential redevelopment that has taken place as the region has urbanized, led by central and local states in close collaboration with large businesses. East Asian academics who have researched gentrification have tended to do so using those classic conceptualizations which originated from Ruth Glass (e.g. Kim and Nam 1998; Qiu 2002; Huang and Yang 2010). This is in contrast to other East Asian urban theory which has highlighted a different production of global modernity – for example the idea of 'Shanghai modern' (Lee 2001) or the notion of Asia as method (Chen 2010). More recent writing about gentrification in East Asia, however, is now much more sensitive to the contextual characteristics of gentrification in the region (see Shin, 2009a; He 2012; and the forthcoming special issue 'Locating Gentrification in East Asia' Shin, Lees and López-Morales 2016). Researchers are now taking heed of, for example, Ma and Wu (2005) who argued a decade ago that seemingly similar forms (like, say, new build gentrification etc.) in Chinese cities may have arisen from different processes and logics thus challenging the Western-centric 'convergence thesis' (pp. 10–12). Shin and Kim (2015) also argue that the urban context of speculative urbanization under the South Korean developmental state has resulted in more endogenous forms of 'new-build' type gentrification from the early 1980s. Indeed, the process of gentrification itself is not so new in East Asia. While the term gentrification may not be widely circulated in the media and public discourses, its ontological presence can be identified through careful scrutiny of how urban neighbourhoods are targeted for capital (re)investment involving displacement of local inhabitants (also see Ley and Teo 2014).

In *South Asia*, the twenty-first century has seen an 'urban turn' (Rao 2006). Researchers are concerned now with how in post-colonial nations, political society finds expression in the city (Chatterjee, 2006). There is a politics of hope, embedded in postcolonial theory, for urban citizenship, that is reflected in the large literature on South Asian cities that looks at the ways in which 'subaltern subjects consent to and participate in projects of urban redevelopment and urban inequality' (Roy 2009: 823). Doshi (2013), for example, looks at the politics of the evicted differently on Mumbai's slum gentrification frontiers. As Roy (2009: 827) states, the subaltern is simultaneously 'strategic and self-exploitative, simultaneously a political agent and a subject of the neoliberal grand slam'.

Traditionally, writings on *African cities* focused on underdevelopment, dependency and informalization. More recently, there have been attempts to understand capitalism in African cities in terms of the circulation of migrants, commodities, etc., creating a rich picture of African urban economies (see Simone 2004). Writings on gentrification have tended to be located in South Africa to date, and it is there that Mbembe and Nuttall (2008) talk about Johannesburg as a key place for the emergence of specifically African forms of modernity and urbanism in which the formal, sanitized Western, white city and the informal, dysfunctional African City are combined.

We have long thought about the *Middle East* as an unstable region of conflict and ethnic nationalisms, of urbicide (Graham 2004), rather than the (re)making of the urban. But more recently, Singerman and Amar (2006) have become a 'Cairo School' to the 'LA School', which studies the modernities and cosmopolitanisms in the globalized Middle East – from shopping malls to gated communities. There are also new articulations of an Arab modernity, what Davis (2006a: 53) calls the 'monstrous caricature of futurism' – where 'the distinctions between the black economy and global finance capital are erased, where city and nature are violently fused, and where the feudalism of an emirate meets up with an open cosmopolitanism' (Roy 2009: 828). Little has been written on gentrification in the Middle East to date (although see a number of chapters in Lees, Shin and López-Morales 2015), urban studies have tended to focus on the architecture and design of new developments/modernities rather than the displacements caused by these gentrifying developments.

These regional snap shots are useful in giving readers a flavour of gentrification and thoughts on it from around the globe. They underline the geography of gentrification (Lees 2000), its contextualities and temporalities, but as the chapters to follow unfold, we put significant meat on the bones of these skeletal musings. The book commands a large body of material, and we have tried hard to make it accessible and interesting to a wide range of readers. The book traverses economic, cultural and theoretical territory, which we have tried to write with a light yet comprehensive hand. Our aim is a nuanced conceptual discussion that critically sets the future agenda for gentrification studies North and South, building from rich, empirical illustrations, an agenda that has social justice at its heart.

2 | New Urbanizations

In the early days of gentrification research, gentrification was often understood as a process that involved a 'back to the city' movement of people, especially the return of middle-class families who were known to have fled inner city areas in order to pursue a suburban way of life based on automobile culture, single family units and the low prices of suburban land that made such a lifestyle more affordable (although we now know that many of these pioneer gentrifiers came from other inner-city areas rather than having their origin in the suburbs). Despite the extended discussions on the causes and outcomes of gentrification, one fixed variable that has long characterized the landscape of gentrification debates is the association of gentrification with inner-city areas, while suburbanization is linked to urban fringes. This tendency is succinctly captured by Tim Butler (2007b) who summarizes that:

> Most accounts . . . have defined the two processes [gentrification and suburbanization] as spatially specific — either to the inner city (in the case of gentrification) or the city fringe (in the case of suburbanization). The problem is that these accounts are rooted in specific moments of the paired process of industrialization and urbanization that took place in the 19th and 20th centuries. (p. 762)

To a large extent, the focus of gentrification researchers on inner-city areas can be said to have resulted from the ways in which the new urban phenomenon of gentrification, originally conceptualized by Ruth Glass, contested the modernist understanding of how cities evolved over time through the rise of suburban affluence at the expense of inner-city dereliction and poverty. In other words, gentrification, which is understood as the repopulation of inner-city neighbourhoods by affluent groups through residential rehabilitation at the expense of displacement of working-class residents, questioned the received wisdom long-preached by the Chicago School of urbanism.

Lees, Slater and Wyly (2008) explain the detrimental impacts of the ways in which the Chicago School has incorporated neoclassical economic theories for explaining urban processes. Having emerged in the early twentieth century with the pioneering publication of the book *The City*, the Chicago School of urbanism provides a modernist perspective on the city, which regards urban conditions as outcomes of rational personal choices based on consumer sovereignty (Dear 2003: 500–2). In particular, the spatial configuration was to take on the form of a series of concentric zones, '[b]ased on assumptions that included a uniform land surface, universal access to a single-centred city, free competition for space, and the notion that development would take place outward from a central core' (Dear 2003: 500). According to the bid-rent theory of land use (see Alonso 1964), suburbanization was explained as the rational outcome of competitive bids by different segments of population, making trade-offs between accessibility and space. That is, suburban neighbourhoods saw the incoming of wealthier families who were in possession of consumption preferences for a more spacious living environment and who had the economic capability to pay for higher transportation costs due to long-distance commuting. In return, inner-city neighbourhoods were understood to have been cramped by poorer families who suffered from high living costs and small and crowded dwelling spaces. Gentrification of inner-city neighbourhoods suggested that the urban evolutionary perspective held by the Chicago School of urbanism was seeing an anomaly, which increasingly commanded its presence in a number of (post-industrial) cities.

The Chicago School of urbanism tried to explain the 'return of people' to inner-city areas without eroding its own belief system. Resorting to the continued use of the bid-rent theory of land use, Schill and Nathan (1983), for instance, explained that gentrification can be a rational response of a select group of upper-income families who outbid all other existing groups near the central city, as they happen to 'value both land and accessibility, and can afford to pay for them both . . . and outbid all other groups for land close to the urban core' (p. 15). In this scenario, the remaining task is to explain why and how the selected group of upper-income families has come to display changes to their consumption preferences. To some extent, the faction of gentrification research that focused on the changing characteristics of individual gentrifiers can be said to have been a response to neoclassical studies on changing preferences for consumption (of land and housing) (see Chapter 3).

Location-wise, as suggested at the beginning of this chapter, gentrification has been associated with inner-city areas or the central city. As such, its decades-long debates bear the footprints of the concentric ring model of the Chicago School or modernist thinking about urban development that assumes the presence of singular centrality in urban spatial configuration. Even if we restrictively define gentrification as an inner-city urban process (though this chapter is to go beyond this narrow definition), gentrification and suburbanization are not necessarily 'a zero sum game either in terms of investment cycles or residential choice' (Butler 2007b: 762). The two processes may happen simultaneously, for instance, due to the expansion of the middle class itself, (domestic and international) migration that add populations to both the city core and the suburbs, or by means of promoting peripheral development such as science parks and gated communities while also investing in inner-city service industries such as financial districts (see Butler 2007b).

With a renewed understanding of urban development and new theoretical insights into urbanization processes, it may be time to revisit these key assumptions in gentrification research. In this chapter, we try to push further Eric Clark's (2005) call for the 'order and simplicity of gentrification', taking into consideration recent debates in critical geography about global suburbanism and planetary or extended urbanization (Keil 2013; Brenner and Schmid 2014; Merrifield 2014). In particular, we pay attention to how urban centrality emerges in multiplicity, going against the modernist assumption of a single core that radiates outwards to conquer rural hinterlands in the process of suburbanization. Andy Merrifield (2014) in his discussion of planetary urbanization refers to the ways in which the centre-periphery relationship is being reproduced across the urban, without being restricted to a singular centrality versus the rest of the urban as the periphery. To a large extent, our approach in this chapter can be said to be a response to this call for removing centre-periphery binary thinking, requiring a more flexible way of conceptualizing how capital reinvestment removes barriers to its operation in the built environment, and hence creates conditions ripe for the class re-making of urban space, including demand creation through speculative activities. Bearing this in mind, we shift our attention to the rest of the world, beyond the usual suspects of gentrification, which presents us with a quite different picture of urbanization and therefore landscape of gentrification.

REFINING A CONCEPT OR STRETCHING IT BEYOND ITS UTILITY?

Theoretical enquiries by means of constant shifting between abstraction and concrete realities inevitably focus on how changing realities influence the refinement, enrichment or abandonment of original concepts. Gentrification has been no exception. It has gone through many rounds of debate on how it can be understood in, and applied to, a particular space and time other than those associated with its original conceptualization. While it is important to pay attention to the complexity of urban processes influenced by a set of on-going and historical contingencies of economic processes, socio-political relations, cultural norms and regulatory systems, it is equally important to avoid excessive conflation of urban processes by paying attention to both 'order and simplicity' (Clark 2005), as we do in our enquiry into the process of gentrification.

The conceptual definition of gentrification has been evolving over time and space, reflecting the expanding epistemological horizon over how the urban is defined and what new trends of urbanization have emerged. Gentrification in its first conceptual appearance was largely referring to the take-over of working-class neighbourhoods by middle- or upper-income households through residential rehabilitation (Glass 1964a). This focus on residential rehabilitation and inner-city location remained for some time. Neil Smith, in his 1982 article on *Gentrification and uneven development*, defined gentrification as:

> the process by which working class residential neighborhoods are rehabilitated by middle class homebuyers, landlords, and professional developers ... The term gentrification expresses the obvious class character of the process and for that reason, although it may not be technically a 'gentry' that moves in but rather middle class white professionals, it is empirically most realistic. (pp. 139–40, n. 1)

For quite some time, residential upgrading or rehabilitation remained as a core characteristic of gentrification, and therefore neighbourhood change through demolition and new build development was differentiated from gentrification. Smith (1982) made this 'theoretical distinction between gentrification and redevelopment', arguing that the latter 'involves not rehabilitation of old structures but the construction of new buildings on previously developed land' (ibid.). For some, the preoccupation with the key characteristics outlined by Ruth Glass (1964a), that is, incremental residential rehabilitation in inner-city working-class neighbourhoods resulting in displacement, has led to overly sensitive approaches to contextualizing gentrification (for example, Maloutas 2011; see a critique of this in the introduction to Lees, Shin and López-Morales 2015). Warning against dogmatic adherence to concepts abstracted out of non-necessary relations, Eric Clark stated in 2005, 'being necessary for explaining a particular case is different from being a necessary relation basic to the wider process. Central location may be one important cause of the process in some cases, but abstracting this relation to define the process leads to a chaotic conception of the process, arbitrarily lumping together centrality with gentrification' (p. 259).

Over time, on-going political and economic restructuring altered the terrain within which gentrification unfolded. Gentrification turned out to be no longer confined to a set of core cities in the global North nor did it restrict itself to housing rehabilitation (see Davidson and Lees 2005, 2010). Developmental projects that sought to rewrite urban landscapes became larger, and so did the institutional support systems that made mega-projects possible in a more market-friendly way. Domestic and transnational financial actors also all geared up so that developers, governments and even individual homebuyers could gain a foothold in the investment frenzy. As property-led regeneration became increasingly pivotal to urban policymaking (Healey et al. 1992), urban redevelopment became the norm, organizing buildings and dwellings into a super-bloc to promote area-based clearances and the reconstruction of upmarket commercial and residential premises. Accordingly, Neil Smith revisited his position on this in his 1996 publication *The new urban frontier*, arguing that it was no longer meaningful to make a distinction between rehabilitation and redevelopment:

> Gentrification is no longer about a narrow and quixotic oddity in the housing market but has become the leading residential edge of a much larger endeavor: the class remake of the central urban landscape. It would be anachronistic now to exclude redevelopment from the rubric of gentrification, to assume that the gentrification of the city was restricted to the recovery of an elegant history in the quaint mews and alleys of old cities, rather than bound up with a larger restructuring. (p. 37)

Five characteristics of urban development influenced the mutation of gentrification, namely 'the transformed role of the state, penetration by global finance, changing levels of political opposition, geographical dispersal, and the sectoral generalization of gentrification' (Smith 2002: 441). In this regard, it was inevitable for gentrification to be redefined as:

> The investment of CAPITAL at the urban center, which is designed to produce space for a more affluent class of people than currently occupied that space . . . Gentrification is quintessentially about urban reinvestment. In addition to residential rehabilitation and redevelopment, it now embraces commercial redevelopment and loft conversions (for residence or office) as part of a wider restricting of urban geographical space. (Smith 2000: 294; original emphasis)

While the definition of gentrification evolved to be more inclusive of redevelopment in addition to rehabilitation, there still remained a central focus on urban centres or inner-city areas, as the two quotes from Neil Smith above clearly show. This focus on inner-city areas or traditional urban cores has continued to dictate the boundary of gentrification debates. For instance, Lees, Slater and Wyly (2008) also define gentrification as 'the transformation of a working-class or vacant area of *the central city* into middle-class residential and/or commercial use' (p. xv; emphasis added); they considered 'other' gentrifications, like rural or suburban gentrification to be related but different. In this book, we move the debate on and argue that this focus on inner-city or singular centrality needs to be questioned further. If we retain the class remaking of urban space and the resulting (in)direct displacement of urban inhabitants (both users and occupants) as the core characteristic of gentrification, as Neil Smith's latest definition suggests, and if we are to take into consideration how urbanization unfolds in those places outside of the usual comfort zone of 'Western' gentrification, we must question the central city as the hegemonic site of gentrification globally. As early as 2005, Eric Clark suggested that there is no reason to confine gentrification to just inner-city areas, if it is to be defined as capital reinvestment that brings about the transformation of neighbourhood characteristics to accommodate more affluent households.

GENTRIFICATION IN THE CONTEXT OF PLANETARY URBANIZATION

As the remit of gentrification research is extended to urbanization outside of Western European and North American regions, gentrification research travels out of its usual comfort zone. The global South experiences urban development that often exceeds neighbourhood scales. Individual projects gentrifying neighbourhoods and urban spaces may come together to create 'metropolitan-scale' gentrification as Shin and Kim's (2015) discussion of the temporal and spatial concentration of urban redevelopment and reconstruction projects in Seoul attests to, a process that can be termed 'mega-gentrification' (see Chapter 7). In Rio de Janeiro, without mentioning the word 'gentrification', Queiroz Ribeiro (2013) claims that the whole urban system of the city has surrendered, since the early 1990s, to a state-led process of service-oriented economic reshaping to reverse the decline of its post-industrial urban economy. In this case, the policy prescription has rested on large-scale housing formalization and social domestication in favelas, especially those considered as the most dangerous and centrally located.

There are also projects at the regional scale. In India, for example, Goldman (2011) traces the development of a regional development strategy that centres on the Mysore-Bangalore Information Corridor in Karnataka State, which includes the construction of an information corridor, an expressway and several brand new townships. The whole development is expected to displace 'more than 200,000 rural people' (Goldman 2011: 566), and to involve land expropriation from villagers, who are compensated poorly, while developers are to benefit from highly elevated ground rents. Rent gap exploitation is at work here (see Chapter 3), in the same manner as it has been carried out in inner-city redevelopment projects, though the process of disinvestment followed by revalorization as discussed by Neil Smith (1979, 1996) has not happened here. Rather it is the expected gains from elevated real estate value, or potential ground rents and the fact that local villagers are rid of their rights to benefit from future development. Indeed, this is the regional blown-up version of what López-Morales (2011) refers to as 'gentrification by ground-rent dispossession' with regard to inner-city redevelopment in Santiago de Chile. Can one say this process conforms to gentrification? If yes, it surely goes beyond the imagination of most gentrification scholars and anti-gentrification activists, and at the same time, presents them with an acute understanding that there is a common thread that connects inner-city gentrification in New York City or London with speculative property development on village lands in India or mainland China. If yes, it surely substantiates the point that gentrification is less a certain predefined urban 'condition' and more a mutating process of urbanization, in which gentrification-led displacement is nothing more than an outcome of already existing unequal power relations within a society, contributing to the intensification of societal polarization. Gentrification is less about predefined loci, morphological attributes and scales, but much more about a process that depends on the contextual reconfiguration of state policies and embedded class and power relations (see Chapter 3). Furthermore, as Lees Slater and Wyly (2008: 138) have noted with regards to rural gentrification studies, 'it is the underlying logic of the process of capital accumulation which unites the urban and rural, and the gentrification of both'.

The brief discussions above prompt us to rethink the nature of urbanization and its meaning in the contemporary world. The starting point is the role of real estate in capitalist societies, building upon Henri Lefebvre's and David Harvey's discussions about the ascendancy of the secondary circuit of capital accumulation. In his thought-provoking book, *The Urban Revolution* (2003), Lefebvre posits:

> I would like to highlight the role played by urbanism and more generally real estate (speculation, construction) in neocapitalist society. Real estate functions as a second sector, a circuit that runs parallel to that of industrial production, which serves the nondurable assets market, or at least those that are less durable than buildings... As the principal circuit – current industrial production and the movable property that results – begins to slow down, capital shifts to the second sector, real estate. (pp. 159–60)

For Lefebvre (2003), the second sector of real estate acts like 'a buffer', a shock-absorber so that surplus capital is channelled into it 'in the event of a depression' (p. 159). Harvey concurs with this line of analysis, arguing that the capitalist urban process is inherently prone to the over-accumulation crisis in the primary circuit of industrial production, which then calls for the switching of capital flows into the secondary circuit of capital that constitutes fixed asset and consumption fund formation (Harvey 1978). The switching of capital occurs in two dimensions, sectoral switching and geographical switching (Harvey 1978: 112–13). The former includes the capital channelling from one circuit to another, for instance, industrial capital channelled into innovation or real estate. Geographical switching entails the transfer of capital from one locale to another, and is particularly pertinent to investment in the built environment due to its immobile nature. But also the switching of capital from the primary circuit to the secondary one is what characterizes the rapidly developing economies of the global South, as many of their policies and economic agents, without necessarily responding to imperialism, reinforce the uneven development of capital across urbanizing spaces as well as regional and national territories.

While the channelling of capital from the primary circuit of industrial production to the secondary circuit of the built environment was thought to be temporary, to address the over-accumulation crisis in industrial production, Lefebvre posits the possibility of the secondary circuit rising above the primary circuit and becoming permanently dominant, stating that '[i]t can even happen that real-estate speculation becomes the principal source for the formation of capital, that is, the realization of surplus value' (Lefebvre 2003: 160). In developing countries experiencing accelerated urbanization and economic development, the need to expand the capacity of infrastructure and housing construction becomes tantamount to an increase in their capacity of industrial production (see Shin 2014a for discussions on China).

When Lefebvre predicts the coming of an 'urban revolution' and hence an 'urban society', he is essentially telling us about the global ascendancy of the secondary circuit of capital, especially real estate based on the capitalist logic of accumulation that subordinates industrial production to the built environment. Like the world market for

Karl Marx, 'the urban' is a vital necessity 'for the reproduction of capitalism on an expanded scale' (Merrifield 2013a: 913). Lefebvre initially observes post-industrial urban societies when he argues that the secondary circuit of the built environment has become 'the mainstay of a global and increasingly planetary urban economy, one of the principal sources of capital investment, and hence over the past 15 to 20 years the medium and product of a worldwide real-estate boom' (Merrifield 2013a: 914), but with its planetary outreach, the speculative frenzy has also usurped the developing world that sees a greater role for speculative land and housing markets in its urbanization processes. It is in this context that we situate the debates on the rise of global gentrifications (see the concluding chapter in Lees, Shin and López-Morales 2015), or neo-Haussmannization according to Merrifield (2013a) that 'integrates financial, corporate and state interests, tears into the globe, sequesters land through forcible slum clearance and eminent domain, valorizing it while banishing former residents to the global hinterlands of post-industrial malaise' (p. 915).

Speculation on real estate has been closely associated with gentrification debates due to its impact on both the production of gentrifiable properties and people's desire to accumulate wealth by investing in real estate properties (see for example, Smith 1996; Ley and Teo 2014). Here, the above Marxist perspective helps us in terms of how the broader structural mechanism of capitalist accumulation and the desire of surplus creation elevates the status of the real estate sector. In Western European and North American countries, the demise of Keynesian welfarism and the rise of neoliberal urbanization have facilitated the hegemonic position of real estate, aided by the financialization of real estate properties, enabling immobile properties to be subject to globalized flexible investment strategies (see Moreno 2013, Weber 2002). In the global South, real estate speculation has also resulted from the pursuit of increasingly mobile capital and professionals, as well as the desire to raise the profile of major urban centres as part of political legitimacy building, often sucking in capital for real estate projects at the expense of more productive uses (see Cain 2014; Goldman 2011; Shin 2014a; Watson 2014). It is in this context, the rise of urbanization at a planetary scale, that we concur with Neil Smith's preposition that the rise of new urbanism in the increasingly neoliberal world has entailed 'the generalization of gentrification as a global strategy' (Smith 2002: 437).

The rise of a generalized gentrification across the globe does not, however, necessarily mean that gentrification gets imported in an imperial fashion, although there is an increasing tendency towards cross-border policy mobility that enables the travelling of 'best practices' (see Chapter 5). As Shin and Kim (2015) explain in relation to the rise of speculative urbanization in the context of South Korea, gentrification as a class-based strategy to advance capitalist accumulation through the spatial fix may be more endogenous and long-established than critics may have thought. Driven by a capitalist urbanization process that sees the rise of the secondary circuit of real estate in conjunction with the primary circuit of industrial production, land and housing markets become attractive destinations for both household and business investments, eventually producing speculative urban policies that depend heavily on market-driven redevelopment to transform disorderly and substandard (from the perspective of authoritarian state elites) urban spaces into organized, middle- or upper-class oriented ones. This speculative nature of urban transformation entices emerging middle classes as strong advocates of urban redevelopment projects (read gentrification), at the same time as disempowering resistance movements against displacement.

In this regard, as Ley and Teo (2014) have succinctly demonstrated in relation to Hong Kong, gentrification may exist and govern the ways in which urban reinvestment changes the social and spatial landscape in existing built-up areas, even if the process is not explained locally by invoking the label 'gentrification' per se. Hong Kong's situation provides an additional dimension in terms of how previous studies of gentrification focused on 'cultural consumption' can be merged more synthetically with attention to fundamental economic processes of accumulation (see Figure 2.1). Ley and Teo (2014) refer to the ways in which collective consumption in the form of mass provision of public housing has long been part of urban policymaking in Hong Kong, and explain how such provision of public housing acts as a buffer to the rise of any meaningful resistance against redevelopment. Furthermore, people's aspirations to accumulate property assets, hence the coining of the term, 'culture of property' has been legitimizing the city-state's drive for city-wide investment and reinvestment in the built environment. That is, 'the tenacious culture of property in Hong Kong has obscured the working of a familiar set of class relations in the housing market, relations satisfactorily described by the concept of gentrification, albeit gentrification in a distinctively East Asian idiom' (Ley and Teo 2014: 1301).

The East Asian experience of speculative urbanization illustrates the importance of the developmental state and their guiding and interventionist role in promoting economic development (and hence their ability to address economic sustenance of the general population to acquire legitimacy for their undemocratic rule). When we shift our attention to a completely different urbanization context such as the Middle East, the underlying logics of capital accumulation and spatial fix through investment in the built environment also provide a landscape amenable to the rise of gentrification, albeit in a Middle-Eastern style (see Chakravarty and Qamhaieh 2015). For instance, Krijnen and De Beukelaer (2015) provide a useful introduction to the situation in Beirut where real estate became an important economic sector in the midst of post-civil war reconstruction in the early 1990s. Low-rise apartment blocks built in the 1950s have been targeted for demolition, to be replaced by high-rise upmarket buildings. The role of

Figure 2.1: Vertical Accumulation in Hong Kong, 2011 (Photograph by Hyun Bang Shin)

civil conflicts and sectarianism, especially in facilitating or confronting gentrification and displacement is pertinent. Politicians engage heavily with the real estate sector, diaspora capital (especially Lebanese expats outside Lebanon working and living in other Middle Eastern countries) buying properties, hence most new flats purchased remain mostly uninhabited. The gentrification is mostly new-build, though some elements of commercial redevelopment through partial rehabilitation have occurred. The Lebanese state, since independence from France in 1943, has never provided welfare; affordable housing provision, if any, occurs through religious or sectarian groups.

Abu Dhabi, along with other UAE cities such as Dubai, has been promoting a spectacular urbanization through mega-projects (see

Figure 2.2: Spectacular urbanization in Abu Dhabi, 2015 (Photograph by Clara Rivas-Alonso)

Figure 2.2) financed by its oil money, and enjoys the reputation as 'a safe haven for investment and an attractive place to live and work' (Chakravarty and Qamhaieh 2015: 60) for large (transnational) corporations and their employees. Nevertheless, the city suffers from a chronic shortage of rental properties especially for the middle- and low-income expat tenants who do not share the fortunes of those highly skilled expats at the other end of the spectrum. In Abu Dhabi, the small circle of master developers that control most of the real estate infrastructure 'are all either majority-owned by parastatals or have such close ties to key members of the ruling family that they are effectively under the government's control' (Davidson 2009: 72). It is also known that stakeholders in real estate corporations are also holding public positions, and there is a blurry boundary between the private sector and the state due to close relationships (see Chakravarty and Qamhaieh 2015). There is a similarity between what happens in Abu Dhabi and elsewhere in the Middle East such as Beirut where there is a stronger fusion between real estate capital and the state, largely due to the fact that (a) Beirut's downtown is depopulated due to the decade-long war; (b) most properties in downtown Beirut were acquired by the developer Solidere, which was established by the former Lebanese prime minister; and therefore, (c) 'political power and private capital investment were held in the same hands', creating more amenable conditions for the wholesale redevelopment of downtown Beirut (see Elshahed 2015: 136).

GLOBAL SUBURBANISMS AND PERIPHERAL URBAN DEVELOPMENT

More recently, there has been new interest in suburbanization, this time guided by scholarly interests in how suburban space has been governed and is reshaped through investments in land and infrastructure often led by a coalition of domestic and transnational elites but also by endogenous inhabitants who adapt to the new suburbanism. Roger Keil (2013) has proclaimed, '[m]uch of what goes for "urbanization" today is not what was seen as such in classical terms of urban extension. Rather, it is now generalized *sub*urbanization' (p. 9; original emphasis). By claiming '*sub*urbanization', Roger Keil is not so much disputing the proliferation of urbanization at a range of geographical scales (in other words, planetary urbanization) as emphasizing the revelation that peripheral areas outside the traditional urban core are where many investment activities are taking place. That is, 'much if not most of what counts as urbanization today is actually peripheral' (Keil 2013: 9). These activities are involving the construction of new-built forms of a diverse nature, ranging from gated communities (neighbourhood scale) to new towns and special economic zones (metropolitan or regional scale).

Selective investment is made in peripheral spaces where the urban meets the rural. Sometimes, leapfrogging takes place to install beachheads of the urban in areas that are predominantly rural by creating zones of exception (Levien 2011; Wu and Zhang 2007). Rural communities become subject to land expropriation and dispossession due to projects that aim to transform such areas into 'world city' locations or to host mega-events to raise the profile of host cities or countries (Goldman 2011; Shin 2012). When an entire rural village becomes subject to land dispossession to produce a new township that is designed to attract elites, professionals and the affluent, is this 'gentrification'? Such urbanization goes beyond the usual geographical scale that gentrification research thus far has attended to, that is, the neighbourhood scale. For instance, in the case of a regional development project to install an IT corridor in Bangalore, India, farmers were forced to sell their lands at a fraction of market-going rates. Another example would be the case of the highly mediatized project of Vila Autódromo (Vainer et al. 2013; Roller 2011) in Rio de Janeiro, located close to the suburban Barra de Tijuca neighbourhood. This case represents one of many similar displacement programmes that have affected thousands of households resulting from a series of investments in Olympic-led infrastructure and mega-transformation of the city (dos Santos Junior and dos Santos 2014). In Goldman's (2011) research, we see the emergence of the IT sector as a beachhead for (justifying) real estate projects, made possible by the joint actions of parastatal agencies, international financial institutions, and transnational policy networks such as international consultancy firms. All these agents and their interests merge together to produce the mega-displacement of villagers whose lands are taken away and villages torn down. The demolition was to clear the way for lucrative real estate projects to build brand new townships as a means to provide homes for national elites and skilled expats. In this process, those dispossessed are rendered vulnerable, while '[l]and speculation and active dispossession inside and surrounding the city of Bangalore is the main business of its government today' (Goldman 2011: 557).

The whole process is a blown-up version of what usually takes place in many new-build gentrification projects that are driven by sub-national government agencies, housing bureaus, (trans)national architectural firms, domestic/global investment capital, and so on. Because compensation is made on the basis of the assessed land value pre-development, the expropriation of villagers' lands in peripheral or rural areas entails the unequal appropriation (dispossession) of ground rents in the form of lost opportunities for villagers and any other previous users of village lands to benefit from the elevated post-development value of village lands. As Goldman (2011) states:

> Under the law of eminent domain, based on the British colonial Land Acquisition Act of 1894, government can acquire land from farmers if it is for a project that is for the 'good of the nation', but it must offer a fair market price (see D'Rozario, n. d.). The Karnataka Industrial Areas Development Board (KIADB), however, offers a relative pittance to the non-elite members of rural communities, exercising its right to choose the depressed rural market price and not the upscale world-city market price as its marker. The difference comprises 'the rent' that shapes and fuels the new urban economy and its governance structure: the black, grey and white markets of real estate brokering and speculation, the mega-deals conjured for new highway, special export zone (SEZ) and township construction, and the new governance system 'managing' these transitional steps toward becoming a world city. (p. 566)

Speculating on land and dispossessing inhabitants therein are at the centre of the transformation of rural peripheral space into a furnace of real estate industries and industrial production. The experience of Mumbai is another testimony to how the secondary circuit of real estate has risen above the primary circuit to bring about the complete urbanization of Mumbai and its adjacent regions. The growing recognition of the role of real estate is also evident in the work of Datta

Figure 2.3: Mega-urbanization in Mumbai, 2010 (Photograph courtesy of Andrew Harris)

(2015) on entrepreneurial city-building in India. She highlights the role of the Indian state in bringing about mega-urbanization in Indian cities (see Figure 2.3), which results in the creation of new avenues of capital accumulation. Obviously the avenue is filled with an Indian version of gentrification or neo-Haussmannization.

Suburban or ex-urban development that installs newly built forms on lands relatively free of legal and socio-political constraints is relatively speaking an attractive option for ruling elites in comparison with development in dense urban cores. Mohamed Elshahed (2015) for instance examines the urban development process in Cairo where suburban urbanization has been a dominant feature of urbanization. Desert land usually owned by the military has been given to private investors through direct sales, resulting in the development of private, gated suburban estates. Despite on-going interest from private investors in exploiting the potential of Cairo's historic downtown, full of heritage sites, gentrification has not taken off there for now, due to extant rent controls as well as legal fights involving owners and inheritors. As Elshahed (2015) summarizes:

> downtown's gentrification potential is limited due to its location in a city where real estate investment is directed towards the desert periphery, where suburban developments are mushrooming ... Potential gentrifiers, young couples, are often unable to borrow from banks (red lining) to buy and restore apartments in the decaying urban core, such as downtown. However, borrowing for the purchase of real estate in new desert communities is facilitated. (p. 125)

However, this does not mean that the historic downtown is to stay as is. In an increasingly booming real estate market, underdevelopment leads to greater development potential. Elshahed (2015) himself hints at latent gentrification in other parts of the urban core outside of the historic downtown area. This is especially with regard to 'the emergence of a rent gap in the inner city resulting from skyrocketing land values with properties of limited investment return' (p. 125), and how the rent gap formation often leads to the acquisition of run-down heritage buildings in pockets of Cairo's urban core (especially neighbourhoods in working-class districts adjacent to the downtown area as well as in the historic core), which are then demolished to give way to tall residential condominiums, a process that is often illegal. This process is further facilitated by the state's effort to exercise the forced eviction of thousands of families, particularly in relation to the creation of a new central business district in the Maspero Triangle north of downtown Cairo (Elshahed 2015: 126).

It is useful to be reminded at this point that the intensification of peripheral urbanization or suburbanization is not to be taken as an alternative to investment in existing urban cores. In fact, the two processes complement each other as part of the profound dynamics of uneven development: '[t]he logic of uneven development is that the development of one area creates barriers to further development, thus leading to an underdevelopment that in turn creates opportunities for a new phase of development' (Smith 1996: 84). Gentrification-induced displacement may also contribute to peripheral urbanization or suburbanization. For instance, as the Anti-Gentrification Handbook for Council Estates in London (London Tenants Federation et al. 2014) shows, peripheral London boroughs turned out to be the destinations of most people displaced from central London. A similar case is witnessed in Santiago de Chile, where low-income displacees from gentrifying central quarters of the city have been pushed to outer suburban areas due to their inability to afford replacement accommodation in their original neighbourhoods or adjacent areas, thus reinforcing the traditional suburbanization of poverty seen for decades in Latin America (López-Morales 2013a).[1]

In an uneven process of development and accumulation, existing urban cores are not to be the exclusive zones of incoming reinvestment, and '[t]he differentiation of the city from the suburbs... will be matched by the continued urbanization of the countryside' (Smith 1996: 85). It is in this context of uneven development and the intensification of peripheral urbanization that we can also revisit existing studies on non-typical gentrification, such as the rural gentrification depicted in the literature on counter-urbanization (see Phillips, 2004). Having mostly grown out of the rural context of the UK, the counter-urbanization literature often examines the characteristics of rural in-migrants, which can be either differentiated from or run parallel to those of urban gentrifiers (see for example, studies by Phillips 1993 and 2002). But Grimsrud (2011) warns against the uncritical application of counter-urbanization that has emerged out of the experiences of core European or American regions, arguing that in Norway, the motivation to initiate urban-rural migration mostly rests on economic or family reasons rather than anti-urban life preferences. Nevertheless, the study of 'wilderness gentrification' by Eliza Darling (2005) in a US-based natural park also reminds us that it is the underlying logic of capital accumulation that gives rise to a localized form of disinvestment that differs from the way it is practised in urban areas but still produces a unique gentrification landscape for both local renters and holiday guests.

NEW URBANISM, RE-URBANIZATION AND GENTRIFICATION

The previous section largely discussed the impact of emerging discourses of planetary urbanization and global suburbanism upon gentrification debates. We now move back to the post-industrial cities of the global North, which were known to have suffered from inner-city depopulation, the emergence of brownfield sites due to deindustrialization, and the financialization of the conventional urban core (especially in some of the leading cities in the global competition for investment). Here, the case of 'new urbanism', popularized in the 1990s and early 2000s in Western European and North American countries, is a useful entry point for its recent influence on real estate markets and gentrification debates. In the context of the detrimental impacts of suburban sprawl, deindustrialization, and to some extent urban shrinkage, 'new urbanism' emerged as a remedy to these negative consequences of contemporary urban development in the post-industrial era of capitalist development. New urbanism called for creating a new sense of community through the promotion of high density construction, new design of buildings and landscapes, and the application of sustainable development initiatives (Calthorpe 1993; see also Moore 2013 for a more critical discussion of New Urbanism and its status as best practice). Inner-city regeneration in the West has increasingly taken the form of new build, often densifying the existing built fabric through demolition and reconstruction of condominiums to maximize returns on investment by developers or individual speculators. Projects under the banner of 'new urbanism' aimed to combine inner-city living with the suburban aesthetic, often branding new residential developments as 'urban villages' or variants. New urbanism projects also frequently targeted derelict or abandoned old industrial sites, changing land uses to accommodate residential projects that usually accommodated young professionals in search of temporary nests before transitioning to a more formal solution. The question was (and still is for some) if this qualifies as gentrification.

In his discussion of the urban regeneration of former industrial land in London's Docklands, Butler (2007b) was more cautious about the use of the term 'gentrification' as a way of understanding the transformation the area had gone through, largely because of the particular aspirations held by the new occupants. Influenced by the rise of new urbanism under the then New Labour government, the London Dockland's regeneration, a project that had been protracted for decades, finally saw its implementation. While the project displayed evident signs of 'regeneration by capital', the new denizens appeared to possess an ethos similar to that of suburban dwellers, that is, 'to be near but not in or of the city', displaying 'a very different profile in terms of gender and family structure – suburbia for singles and empty nesters, as it were' (Butler 2007b: 777).

Butler's focus on the habitus of gentrifiers (new occupants or denizens as in the re-urbanization discourse) obviously had an effect on how gentrification was understood. In the socio-political framing of contemporary cities centred on urban decline and poverty, especially inner-city areas, public policies to bring about urban revitalization and urban renaissance are seen as positive measures that produce new-build residential developments, coined 're-urbanization' by some, rather than 'new-build gentrification' – see for instance, Boddy's (2007) discussion of the inner-city redevelopment of Bristol, which involved the conversion of largely commercial properties into new-build residential properties to accommodate young professionals in transition, residing largely in buy-to-let properties. The absence of clear signs of direct displacement of existing resident populations by the new denizens nor overt political contention over the new developments led Boddy (2007) to reject use of the term 'gentrification' and to opt for the more politically neutral term 're-urbanization' (see Davidson and Lees 2005 and Slater 2006 for a critique). This conclusion was despite the clear signs of speculation by absentee owners (in the form of buy-to-let investments in new developments) and developers, and the on-going classic gentrification in adjacent neighbourhoods, which may have been concluded by the class remaking of the inner-city areas.

A number of problematic positions can be identified in the above re-urbanization arguments. These include the labelling of buy-to-let buyers as investors despite clear signs of their speculative behaviour. Furthermore, a restricted understanding of displacement is apparent in the writings. The re-urbanization thesis refers to the absence of displacement as key evidence in its rejection of gentrification, but for Boddy, displacement is limited to the absence of direct last-resident displacement, which is only one of the four dimensions in Marcuse's (1985a) insightful conceptualization of displacement (the other three are chain displacement, exclusionary displacement and displacement pressure). There is lack of account for the displacement of existing retailers or work places as well as workers who used to have some degree of affiliation and sentiment to these places. The claim of re-urbanization based on a lack of physical displacement of a working-class population is repudiated by Davidson and Lees (2010: 408) who argue that new-build residential redevelopment accompanies 'the alteration of the class-based nature of the wider neighbourhoods'. In particular, they call for a more nuanced understanding of displacement 'as a complex set of (placed-based) processes that are spatially and temporally variable' (Davidson and Lees 2010: 400) rather than reducing it 'to the brief moment in time where a particular resident is forced/coerced out of their home/neighbourhood' (ibid.). Invoking both Marcuse's typology of displacement and also Yi Fu Tuan's distinction between space and place, Davidson and Lees (2010) argue that 'displacement is much more than *the* moment of spatial dislocation' (p. 402; original emphasis), and that 'A phenomenological reading of dis*place*ment is a powerful critique of the positivistic tendencies in theses on replacement; it means analysing not the spatial fact or moment of displacement, rather the "structures of feeling" and "loss of sense of place" associated with dis*place*ment' (p. 403; original emphasis). In this respect, the class remake of urban space through capital reinvestment such as those seen in inner-city regeneration led by the (new urbanist) rhetoric of social mix or urban renaissance (see Bridge, Butler and Lees 2011) actually contributes to the displacement of working-class and poor residents and increases their alienation from the very place they have long been attached to. We expect to see greater attention to the phenomenological aspects of displacement in future studies on gentrification, and on the sentiments of inhabitants (users and owners alike) due to the loss of home and work places in gentrification processes.

CONCLUSIONS

Earlier in this chapter, we discussed how gentrification debates have long been preoccupied with a geographical focus on inner-city areas, even if there were attempts to reach out to the suburban and rural instances of mutating gentrifications (see Lees, Slater and Wyly 2008: 135–8 and Charles 2011). To some extent, the earlier focus on inner-city areas or singular centrality reflects the legacy of the modernist thinking of how cities have evolved over time. More recent innovative and path-breaking thinking of the urban and urbanization at multiple scales has steered our attention to a rethinking of centre-periphery relationships and also the diverse nature of urban development that results in multiple centralities. In short, we believe that a certain specific geographical locus (the inner city) should not be taken as a necessary condition for gentrification. Neither should gentrification be taken as a certain predefined condition (old architecture) and scale (e.g. local, neighbourhood).

Cities aspire to become world cities, and this aspiration is contagious, inflecting other regional cities within a country to emulate the 'worlding' of more influential cities (Roy and Ong 2011). One of the major outcomes of this is the heavy speculation on real estate, resulting in the rise of speculative urbanism (Goldman 2011; Shin 2014a). As land values rise, capital finds real estate more seductive and stays there. Not all displacement is gentrification induced (see Chapter 7 on development-induced displacement for example). But it is important to acknowledge the positioning of gentrification as a process in condensed urbanization. For Merrifield (2013b), however, it is neo-Haussmannization that sweeps the urbanizing landscape, turning each land parcel and landed property into virtual commodities floating in the air, to be snatched by speculators:

> the world now is a project of neo-Haussmannization. In Haussmann's Paris we saw the centre being taken over by the bourgeois and the poor people being dispatched to the periphery, to the *banlieue*, and the centre was commandeered by the rich, and the poor were then displaced to the periphery, particularly to the northeast of Paris. Now, what I want people to do now is to imagine this not occurring in one city, but to see the whole urban fabric as being put in place through a process of neo-Haussmannization, whereby centres and peripheries are everywhere. (p. 13)

The onset of neo-Haussmannization is perhaps the beginning of a planetary gentrification that no longer makes a distinction between central and peripheral locations, between neighbourhood-scale and metropolitan-scale real estate projects, between incremental upgrading and wholesale redevelopment, between physical displacement and phenomenological displacement, and between residential and commercial make-over. The underlying commonality is the logic of capital accumulation, especially the ascendancy of the secondary circuit of real estate, which enables multiple forms of gentrification, thus gentrifications in a plural sense, taking place around the world in a variegated way. The place specificities of this process are important, as many critics highlight in terms of emphasizing the contextual understanding of concept and phenomenon (Lees, 2000), but an equally important emphasis is on the 'order and simplicity of gentrification', which allows us to move from localized neighbourhood transformation to the greater picture of actually existing capitalist accumulation that brings together not only (domestic and transnational) gentrifiers but also urban social movements that challenge the existing social, economic and political order.

But when gentrification expands at the planetary scale, what is it going to become? Is it going to experience a qualitative transformation, becoming something else? Ruth Glass did not have the chance to observe global gentrifications – neighbourhoods were the units of gentrification in the 1960s and 1970s, although she was very much aware of the importance of scalar thinking (see Chapter 1) in that neighbourhood transformation was not understood as isolated from wider metropolitan changes. Many research projects on gentrification have been bounded by the traditional notion of the city, inward looking, without making connections with the rest of the city, region and the world. Emphasizing the importance of understanding how urbanization unfolds in peripheral suburban regions, Keil (2013: 10) states:

> There is much blurring and bleeding among and between the different world regions. In a post-colonial, post-suburban world, the forms, functions, relations, etc. of one suburban tradition get easily merged,

refracted, and fully displaced in and by others elsewhere, near or far. Our optics has changed accordingly and we have collectively been challenged to abandon historically privileged sites for observing urbanization.

Gentrification research has also gone through various challenges over the years in abandoning its classical assumptions, such as a focus on inner-city areas, residential landscapes and rehabilitation. But it may be necessary now to push further this theoretical re-orientation in order to steer research efforts towards integration with new understandings of the urbanization processes that are unfolding, as we speak, at the planetary scale.

3 | New Economics

There seems to be a correlation between gentrification in certain places of the global South and rapidly developing, urbanizing economies. The gentrification of cities at the 'fringes' of the capitalist world shows us several ways in which neoliberalism can unfold; the way the local dominant classes, sometimes global, deploy their power spatially, and the way states intervene to realize their goals. We understand neoliberalism here not as a colonizing force that renders local relations to succumb to transnational elites, but as a contextual force that interacts with local power relations to produce variegated capitalist relations in each locality. For instance, the East Asian developmental states (Taiwan, South Korea, Singapore and Japan) were often known as 'vassal' states, having huge dependence on the US during the Cold War era of the twentieth century, but the region did not really go through neoliberalization as such during this period and these countries' policies were developing and maturing over time (see Park, Hill and Saito 2012; Tsai 2001; Yeung 2000). It should be acknowledged that neoliberalism has been evolving and 'travelling' at an uneven pace, which also in turn affects how certain 'neoliberal' policies have travelled across the world. Capitalist relations at multiple geographical scales are going to be inherently uneven across national borders and often within borders as well (see for example Zhang and Peck 2014), and gentrification is a spatial process that polarizes those relations even more. From this point of view, to claim a singular 'form' of gentrification makes no sense at the planetary scale.

In this chapter, we deal with six economic aspects to see what is essentially new about planetary gentrification. First and briefly, gentrification can be a device for producing capital switching from the primary circuit of industrial production to the secondary circuit of the built environment, but now on a global scale. Second, gentrification is a massive form of creative destruction of capital fixed in the built environment replacing outmoded fabric with newer infrastructure; this multiplies the amount of capital including speculation on real estate, always in search of higher rates of return in the next cycle of accumulation. A third aspect is the production, uneven accumulation, and disputes over ground rent in redeveloping urban areas, a process that Slater (2015) has also called 'planetary'. Some time ago rent gap theory was depicted as over deterministic, but it has recently seen a revival as it has proven useful in addressing the essentially neoliberal inequalities between economic forces and social agents in many rapidly transforming cities, through the complex interplay between national, metropolitan, urban and neighbourhood levels. The rent gap also helps measure gentrification-led displacement, for the private capture of ground rent (López-Morales 2011, 2013a; Shin 2009a,b; also Slater 2006, and Marcuse 1985a) always has a class-monopoly nature (Jaramillo 2008). There is also an ongoing comparative discussion being held in many Latin American countries, about land value capture by national states, as part of socially redistributive agendas of the rent gap (see Smolka 2013). After the 1990s, debates over the rent gap faded away, and more recently, despite Slater's (2006) argument that gentrification researchers need to look more at the displacement of low-income populations, there have been very few attempts to link gentrification economics with displacement and exclusion, and to address the seemingly crucial question of who really captures the ground rent when an area comes to be redeveloped, and what happens when the rent gap levels are so unequally captured that they effectively dispossess land value from local populations who previously had claims on that land; even if they chose to sell, or even if they wanted to stay. The latter 'effects' were not clearly seen in the 1980s and 1990s when rent gap theorizations were at their height in the global North; but we argue now that the conflicted nature of the land-economic side of gentrification is one of its main common 'planetary' characteristics.

There are also other forms of class-led appropriation of urban resources – one is 'spatial capital' (Rerat and Lees 2011), the way transport-oriented public policies transform spatial opportunities in the city, with the explicit (or not explicit) aim of spatial appropriation by upper-income social groups (Blanco et al. 2014); this is a fourth aspect we address here. A fifth aspect is the contemporary forms of public and private sector involvement in inner city *favela, colonia, gecekondu* redevelopment in the capitalist world, in what has come to be called 'urban entrepreneurialism' (Harvey 1989a). Given the increasingly intense scholarly and political interest in the phenomenon of urban 'policy mobilities' (McCann and Ward 2010; Robinson 2011a; see Chapter 5), we explore the intertwined roles of the state and holders of economic capital in the production, distribution, and representation of urban exclusion and segregation, and the roles that gentrification plays in the production of these effects (we also discuss transit-oriented development in Chapter 6).

A sixth, additional fundamental aspect is whether there is a new economy of gentrification in the global South and North that essentially differs from the post-war economy of gentrification in the North (as the dominating narrative of gentrification in the literature). In the final section we address the impacts gentrification has generated in the post-socialist world and also Tiger economies (South Korea, Taiwan and Hong Kong) and the industrialized capitalist economies different to transitional economies, such as China and Vietnam. The greater integration with global investment capital flows and the growing importance of real estate interests in cities in Brazil, Chile, Mexico, and so on, suggests that there exists a frequently conflictual interaction between 'traditional' urban space (often characterized by slums and decaying inner-city places) and emerging 'gentrified' urban space that caters for the needs of the new rich and international visitors. Finally, we move back to the global North to see what the debt crisis has meant for gentrification in the US.

GENTRIFICATION AND THE GLOBAL ECONOMY

Inspired by Marx and Engels' *Communist Manifesto*, Schumpeter (1976) synthesized capitalism as a 'process of creative destruction' and claimed that capitalism revolutionizes its economic structures from 'within', incessantly creating a new one and destroying the old one. Capitalists destroy and replace their fixed capital, means of production, physical and social infrastructures, according to the pace of new technological advances, in order to remain competitive. This is precisely the force that sustains the capitalist city as a machine of production and competition, and gentrification as a way to open up spaces for new rounds of profit and accumulation, especially when the scale of gentrification involves comprehensive state policies for city transformation and the involvement of large-scale redevelopers and financiers. This is happening not only in most of the countries of the global South but also now in the US (Mueller 2014; we come back to this case later) and the UK (Davidson and Lees 2010). Obviously, there has been a leap from the period of 'Haussmannization' and the *Communist Manifesto* to contemporary capitalism and neo-Haussmannization. The latter now spreads as 'a similar process that integrates financial, corporate and state interests, tears into the globe, sequesters land through forcible slum clearance and eminent domain, valorizing it while banishing former residents to the global hinterlands of post-industrial malaise' (Merrifield 2013a: 915).

Drawing on these ideas, Schumpeter agreed with Marx that capitalism would ultimately find its definitive exhaustion, but the fact is that capitalism so far has been able to mutate and evolve into new forms of industrial organization. All over its development, capitalism has regularly fallen into major crises that drive capitalists to switch from the primary circuit and invest in the secondary circuit, in forms of fixed capital and, more specifically, the built environment and urban space. David Harvey (1973, 1982, 2010a) has demonstrated how city space performs a greater role in this process of capital amplification, calling this process 'capital switching'. For instance, the subprime mortgage financial crisis that exploded first in 2008 in the US and the UK, and was apparent in the rest of Western Europe (where we could see its devastating consequences for low-income, elderly and ethnic minority populations) is still a vivid example of this. It was capital that, for more than a decade, had regularly switched from the secondary circuit into the speculative financial markets of mortgage-backed securities. And then, after the crisis, this money returned from the financial sphere to the secondary circuit to be converted into real estate capital again by the 'hyper-commodification' of urban land and other basic social necessities like housing, transport access, public space, and other public goods like healthcare, education, and even water and sewage disposal (Brenner, Marcuse and Mayer 2012). Another good example is the so-called Asian Crisis of the late 1990s that shocked many Asian economies and then hit the Latin American region for five years or so. In this latter example, the secondary circuit of capital accumulation gained particular importance due to its function in absorbing shocks generated from the main production circuit (Shin and Kim 2015; Shin 2014a); there, the secondary circuit was a means to boost the economy and re-establish economic growth, a classic strategy also evident in the New Deal projects during the Great Depression in the US and later elsewhere in the Western world.

In contrast with other more volatile economic sectors, urban space is an efficient form of capital fixation, allowing the processes of accumulation to work within certain levels of stability in profit rates (e.g. the constant rent produced by a residential building). However, cities can also be destroyed in order to clear space for new accumulation. The re-organization of city space is thus not only an expression of the globally induced crises of capital accumulation, but a device for managing them locally (Harvey 1973). Outmoded spaces act as barriers for a more rapid renovation and need to be removed somehow because urban space never changes as fast as the pace of economic development demands. Or, as Rachel Weber (2002: 519) says, the 'accumulation process experiences uncomfortable friction when capital (i.e. "value in motion") is trapped in steel beams and concrete'.

One of the deepest and most pervasive cases of global-scale creative destruction generated by capital switching was the de-industrialization of industrial cities in advanced capitalist nations, namely the UK and the Midwest of the USA, from the early 1970s and for over a decade afterwards (Hall 1999). Post-Fordist flexible production techniques replaced industrial structures and infrastructures developed under the Fordist regime, while once vibrant industrial regions were emptied and replaced by urban economies largely determined by the secondary circuit, where the exploitation of land for real estate purposes was probably the main device for capital generation. This is what Lefebvre (2003) called the 'Urban Revolution'.

Another much less-known example is the first experiment in global neoliberalization, in Chile, in the early 1970s. Neoliberal reconfiguration there was aimed at dismantling the institutional apparatus and productive capacity established during a previous Keynesian regime of industrialization, so as to remove as many barriers as possible to international trade, liberalization of domestic capital markets and openness to external financial markets (Gatica 1989). The consequence was a considerable drop in the share of manufacturing from 26.3 per cent of

GDP in 1960 to 21.6 per cent in 1980, and a fall in manufacturing jobs from 40.7 per cent of the national total in 1960 to 16.5 per cent in 1979. Whilst during the previous Keynesian regime, between 40 and 50 per cent of the workforce living in the most precarious enclaves belonged to the industrial proletariat (Castells 1974), during the neo-liberal dictatorship, from 1975 onwards, this number decreased to 20 per cent, whereas workers became precariously self-employed (35 per cent), and the level of unemployment among the lower-income sector increased from 18 per cent in 1971 to 35 per cent in 1984 (Chateau and Pozo 1987). A massive switch of capital was made possible in 1976 through the abolition of taxes on underdeveloped land, lower taxes on land transactions, liquidation of state-owned reserves of urban land, delivery of property land titles to more than 100,000 households in Santiago alone (for a population at the time of 4 million) that now faced new tax bills, the eviction of residents from (now) illegal *campamentos* in areas of highest land value, new laws that implied the elimination of most of the restrictions on urban expansion, and the dismantling of the existing state-built social housing apparatus via setting up a *laissez faire* land market with open access to private financers, realtors and developers. This was followed by the easy flow of international credit as the financial sector saw a big opportunity to invest the now idle formerly industrial capital into the land and property markets.

A more recent sociological example of post-crisis urban strategy for macro- and micro-economic recovery is given by Alexandri (2015) in Athens, where gentrification is seen by the state apparatuses as a 'noble' contribution from the private sector to society, when in reality gentrification is a class-war against illegal immigrants, drug users and the homeless that use the devalued areas of the city now targeted for redevelopment, amid increasing rates of unemployment. The flows of real estate capital in Greece have less to do with the real needs of people (for housing, open spaces, amenities, social reproduction, etc.) and more with the needs of private capital to stabilize, reproduce and expand. This case shows how capital switching is precisely what Marx (1973, Notebook VI) defined as one of the immanent means in capitalist production to check the fall of the rate of profit (crisis) and accelerate accumulation of capital-value through the speculative creation of new capital, at any political or social cost.

What early 1970s Chile and post-2008 Greece show is that in reality no *laissez faire* exists in neoliberalism, but a market protected by continuous flows of state subsidies and market-friendly land flexible regulations. Both neoliberal authoritarian states are systems of rapid capitalization via land rent private accumulation, thus resulting in privately led urban market expansion. Both cases show how urban space is not only a condition for production (e.g. hosting manufacturing) but the element for production itself, subject to exploitation and hence accumulated without intervention of the primary circuits of capital, or capital flowing into productive spheres. In short, with capital switching into secondary circuits, land rent, instead of profit extracted from labour exploitation, is exacerbated as a profitable commodity *per se* (Harvey 1989a). One exception could be current China, where the switching of capital from manufacturing into land/real estate circuits has not happened after any economic crisis of overaccumulation or political shock, but strives to accelerate *both* urbanization and industrialization at the same time; as an accelerated growth has been sustained by the national regime for at least two decades (see Chapter 7).

GENTRIFICATION AND LAND RENT

The devaluation of capital that has been previously fixed in specific parcels of urban land, or the valuation of other areas that become targeted for new rounds of investment leads to a situation where the ground rent capitalized under current land uses is substantially lower than the ground rent that could potentially be capitalized on if the land uses were to change. Neil Smith (1979) called this process a 'rent gap', namely when redevelopment becomes sufficiently profitable, capital begins to flow back into disinvested land parcels, the redevelopment of outmoded fabric becomes the dominant form of urbanization, and then substantial fortunes can be made, often at the expense of low-income people currently occupying that temporarily devalued land and subject to dire displacement.

We have witnessed for some time now how an array of policy makers and top-ranked officials from all around the world have justified redevelopment-generated displacement as an irrelevant negative side of a largely more positive gentrification coin (see Slater 2006). But in fact, displacement is a substantial part of gentrification and in this chapter we substantiate this affirmation, for rent gap theory provides a powerful means to assess what has come to be called 'ground rent dispossession' or a structural form of social displacement and exclusion (López-Morales 2011, 2013a). Neil Smith (1979,1996) provided an explanatory model for both inner city decline and 'regeneration', where the rent gap was (after its 'realization') integrally set within the logic of the circulation of capital within the secondary circuit and monopolized (see Lees, Slater and Wyly 2008, for a detailed review).

In the global North of the 1970s and 1980s, the declining industrial inner city was an effect of the movement of capital to the suburbs where higher returns were more easily attainable. The same happened in peripheral contexts like neoliberalizing Chile in the 1970s. Thus, a combination of concerted disinvestment by investors in the inner city, due to its high risk and low rates of return, triggered a long period of deterioration and lack of new capital in these areas. After decades of sustained suburbanization, a valley in the ground rent curve deepened in the inner city due to a continued lack of productive local capital investment. Over time, a rent gap appeared. This gap is the disparity between the 'capitalized ground rent' (CGR, rent attracted by a piece of land), devalued by the current dilapidated use of land, and a 'potential ground rent' (PGR), increased by the new improvements in the surrounding area. So, PGR implies the 'highest and best use', or at least higher and better use given the central location of inner-city space (Smith 1996). However, its realization can come exclusively from the development that involves an intensity of fixed capital investment designed to accommodate this potential use.

At the present time, such a post-industrial situation does not exist any more. So we need to ask first, who possesses the means of making the potential ground rent realizable, especially when potential ground rents are much higher everywhere due to the increased (local and global) demand for urban land amid a far more globalized urban economy than in the 1980s. For the appropriation of the rent gap, the state, private owners and investors also play new roles; the former more decisively (than ever) to create the economic, legal, and administrative framework; the latter still responding to its private interests over land rent accumulation but now they are more mobile and transnational. Although there are still few empirical investigations of rent gap theory in countries beyond the global North, the global South world is currently showing a growing array of cases.

TOWARDS A COMPARATIVE URBANISM OF PLANETARY RENT GAPS

Since Smith's (1979) theorization, rent gap theory has been a key feature not only in the economic causes of gentrification but also in observations of conflictual land and housing markets all over the world (Slater 2015). To some degree, this confirms the assertions by Lees (2000), Smith (2002), Lees (2012, 2014a) and Lees et al. (2015) that gentrification has gone global insofar as capital has expanded in search of new spaces of profit in tandem with entrepreneurial policy transfers (see Chapter 5); but we need to more specifically scrutinize how this happens, because as we said, gentrification at the planetary scale is not a 'force' that diffuses from North to South, but an outcome of the interplays between global and local politico-economic forces, intertwined with an ample array of different institutional arrangements that characterize the currently hyper-connected capitalist world.

Sýkora (1993, 1996), in Prague, at the dawn of post-socialist market transition, was one of the first (outside the Anglo-American context) to document an emerging land price gradient, showing 'functional gaps' generated by the underutilization of available land and public buildings under the socialist command economy (this case was extensively documented in Lees, Slater and Wyly, 2008). More recently, Wright (2014) has produced a quite different, feminist analysis, of the *Centro Histórico* in Ciudad Juárez, on the Mexico-USA frontier, amid the transcontinental drugs war and the failure of the states of Mexico and the US to put an end to the warfare (especially the carnage of lower-income women) and tackle the unprecedented role that drug trafficking has in the GDP of those economies. Wright found in Juárez that the rent gap theory was highly applicable in accounting for a situation whereby ruling elites and the city government attempted a redevelopment plan, a strategy that comprised denigrating the lives and spaces of residence and work of lower-income women and children, and the eviction (by using eminent domain) of any kind of business that did not match with the entrepreneurial state-led development plan. The aim was to reduce the capitalized ground rent, enlarge the rent gap, and ultimately close it by wiping out the area and re-establishing it as a place for upstanding households.

Indeed, new issues have arisen from new rent gap analyses that are different to the traditional narratives of the 1980s in the global North (Smith 1979, 1996 and Clark 1987), which had assumed it was landlords who exclusively captured rent. But we now realize that capitalized ground rent can be captured by multiple actors, for example, land owners (who can be low-income residents), developers (usually the dominant stakeholder), and even transnational classes operating in a local territory as Sigler and Wachsmuth (2015) have shown in Panama. In cases where zoning laws are flexible and attractive for redevelopers, and after decades of state policies of mass land titling, like in Chile and Mexico, rent gap analysis makes visible a huge difference in the ground rent levels obtained by redevelopers, who build at the highest land use permitted by local building codes. Therefore, in urban economies based on high rates of petty land-ownership (paradoxically, this is an outcome of many neoliberal policy prescriptions), when redevelopment frenzies start, land-owner households who receive a lower ground rent when they sell the land they inhabit, especially when there is more than one family, are doomed to be excluded from the housing market as they cannot afford to purchase any renovated high-rise residence. When those households come to sell their land to real estate redevelopers, the price does not come from competitive bidding but a sort of monopsony (where one buyer is faced with several sellers) that reduces the cash value for their land and limits their post-occupancy options. This has happened in Chile (López-Morales, 2011, 2013a), Mexico City (Delgadillo, 2014) and Seoul (Shin 2009a), where local owner-occupiers have been co-opted by joining the property-based interests to claim a stake on ground rents. The redevelopment of *Casco Antiguo* of Panama City (Sigler and Wachsmuth 2015; see Figure 3.1) also shows a form of 'transnational gentrification' where the leisure-driven migration of a transnational class with considerably higher purchasing power than the original population, supplied the demand for neighbourhood reinvestment schemes thus increasing the potential ground rent to a point that existing local demand was not allowed an opportunity for profit. In fact, this latter study offers a quite imaginative way to understand the rent gap from a global – even imperialistic – perspective.

The rent gap research by López-Morales (2010, 2011, 2013a,b) in Santiago, Chile, has shown that conflict emerges when the portion of

Figure 3.1: Gentrified Casco Antiguo, Panama City (Photograph courtesy of Thomas Sigler)

the rent gap paid to the original petty landowners for their land is not high enough for them to find replacement accommodation in the area. Low-income petty landowners also lack bargaining power for they usually have limited education and less access to legal means, and because developers acquire land plots in advance in order to avoid the entry of other competitors, and therefore can exert extra pressure on the remaining landowners in the same block (this can be called 'blockbusting' for the reasons explained below). In the poorest, most derelict quarters of Santiago, where original resident households range from middle- to lower-income, or where there is a considerable presence of extremely low-income and immigrant tenants in still non-gentrified residences, redevelopment is more harsh on local people.

In addition, the nature and tremendous power of potential ground rents to attract capital and political forces together has only recently been discussed in any detail. For instance, if Smith (1979) and Clark (1987) saw the concerted devaluation of properties and land by local realtors and financiers as a way of enlarging the rent gap, more recently Hackworth (2007) defines post-recessionary gentrification in the US and Canada as a moment where real estate markets rely less on reduced capitalized ground rents, and far more on state-led land and building regulations as 'amplifiers' of potential ground rents. In this vein, López-Morales, Gasic and Meza (2012) have critically examined the role of national-level state policies in Chile in the physical transformation of neighbourhoods through the amplification of Floor Area Ratios (FARs), namely the ratio of a building's total floor area to the size of the land upon which it is built. The same has been done by Shin (2009a, b) in China and South Korea, and by Sandroni (2011) in Brazil, where land use re-zoning and FARs have also been identified as the main drivers of land rent increases and their uneven social distribution.

Shin's (2009a) discussion of the role of the state in rent gap formation amid substandard settlements in Seoul, South Korea, provides a useful perspective on how the long-term designation of substandard settlements as redevelopment districts acts as redlining (see also Lees 2014b,c, on public housing estates in London). Together with a lack of *de sure* property-ownership, state-imposed conditions present disincentives for any potential investors who consider long-term investment in properties in such informal settlements, thus preventing the arrival of individual 'gentrifiers' until uncertainties are addressed. But while disinvestment leads to the creation of a rent gap in such substandard settlements, the gap is further widened due to the increasing disparity between these settlements and adjacent urban districts that experience real estate booms in times of highly speculative urban development (see also Shin and Kim 2015). Nevertheless, the rent gap closure, and hence gentrification, is only made possible by the

Figure 3.2: Devalued properties with high-rise building in the background in gentrifying Santiago, 2014 (Photograph courtesy of Daniel Meza)

intervention of concerted efforts by external property-based interests who also co-opt poorer owner-occupiers who are given a share of ground rents.

In Seoul, Shin and Kim (2015) further confirm that the endogenous dynamics of urban redevelopment have provided fertile ground for the rise of 'new-build' gentrification from the early 1980s, dictating the course of urban spatial restructuring that has been carried out in a highly speculative manner. This was made possible initially by one of the world's most aggressive residential renewal programmes, namely the 'joint redevelopment programme' (JRP), and subsequent redevelopment programmes that have roots in the JRP, all of which have been designed to maximize landlord profits rather than improving the

Figure 3.3: Demolition in a redevelopment project site in Seoul, 2013 (Photograph by Hyun Bang Shin)

housing welfare of low-income residents who have been displaced from their neighbourhoods (Ha 2015) (see Figure 3.3). Other new cases include piecemeal operations, like in the vast Palermo area in Buenos Aires (Herzer 2008), but also the (often foreign) speculation in, and capitalization on, land in the form of mega-events, large construction projects, urban 'regeneration' schemes and assorted 'growth machine' agendas (Porter 2009; Raco 2012; Shin and Li 2013). Shin and Kim (2015) see Seoul's new-build gentrification not just as a replica of the global North's new-build gentrification, but as revealing distinctive characteristics given the strong influence of the neoliberalizing state, as the process of gentrification in Seoul itself could not have been consolidated without the presence of these large conglomerates as the major partners of the Korean developmental apparatus, something that was not considered in earlier gentrification literatures.

Linking gentrification with the local geography of power relations is also key to seeing the capacity of the extremely powerful real estate and construction sectors in the emerging national economies of the global South that permeate national and local-level states and have advocated a series of entrepreneurial policy prescriptions and implementations in central areas, from Beijing and Seoul (Shin 2009a,b) to Mexico City (Delgadillo 2014) and Buenos Aires (Herzer 2008; Herzer et al. 2015). As cases around the world show, behind the supposedly positive effects of gentrification, lie powerful private economic forces and institutional arrangements that speculate with urban land, create and capitalize ground rent increases by putting land to its 'highest and best uses', and supply increasingly expensive dwellings that – if public policies on housing and land were genuinely redistributive – should instead be aimed at low- and middle-income users. In this way a growing problem of housing unaffordability in central locations globally is (being) created. In fact, the implementation of an increased number of public subsidies aimed at the acquisition of middle-income strata housing, developer-friendly land use zoning at local and metropolitan levels, as well as a number of tax exemptions that benefit large-scale developers, are not natural processes but are political strategies of gearing urban land markets towards generating considerable revenues for the private development and finance sectors.

There are also different forms of blockbusting and illegitimate pressure by landowners on owner-residents to sell and leave, for instance the negative impact of condo construction on adjacent properties (Gaffney forthcoming), and the role of financial institutions and especially the state in redlining certain areas (López-Morales et al. 2014), generating abandonment and speculative deterioration of properties. Economic practices vary from case to case but the speculative landed developer's interests in cities is roughly always the same: how to buy property as cheaply as possible and sell it at the highest price.

STIGMA AND DEVALUATION

The negative way in which certain parts of cities in the global North and South, targeted for renovation, are portrayed by politicians, policy-makers and/or think-tanks is critically important to debates about the rent gap (Slater 2015). A growing body of work points to a direct relationship between the defamation of place and the process of gentrification (Wacquant 2007; Wacquant et al. 2014; Lees 2014b,c), but so far there is no evidence that territorial stigmatization intensifies the rent gap, and further investigations are needed to understand how the theory might be recalibrated to account for the pressing issue of the symbolic defamation of space (Slater 2015). This recent debate very much recalls past discussions on the discourses of blighting. Rachel Weber (2002) suggested that the built environment experiences a greater degree of flexibility and is receptive to real estate capital investment in times of neoliberal urbanization because 'discursive practices that stigmatize properties targeted for demolition and redevelopment have become increasingly neoliberal' (p. 519). She argued that obsolescence has become 'a neoliberal alibi for creative destruction', which concentrates on areas with the highest return on investment in a market that has been increasingly entwined with global financial capital (ibid. 532). To aid and perpetuate this process the local state 'operates through decentralized partnerships with real-estate capitalists, and what remains of the local state structure has been refashioned to resemble the private sector, with an emphasis on customer service, speed, and entrepreneurialism' (ibid. 531). In this regard, both stigmatizing places and, in parallel or soon after, improving the 'quality' of the built environment has been one of the main urban accumulation strategies, increasingly adopted by many governments at both national and sub-national scales.

The contradictions between the needs to destroy fixed capital (or consumption funds) and increase the rate of return by land rent exploitation, finds one of its best examples in contemporary Hong Kong. In a recent study, La Grange and Pretorius (2014) see how much of the city's residential buildings in the inner area are depicted as old and obsolete, but at the same time what is really striking is that they provide affordable, well-located housing for lower-income and disadvantaged groups and small-scale commercial clusters. The gap between current and future potential landed profits is evident in Hong Kong, but this is not the result of economic decline (as we see in the 'post-crisis' section of this chapter, next) but rather of formidable frictions that make land assembly and vacant possession of buildings, in a particularly hyper-dense urban morphology difficult and where redevelopment needs to draw on strong policy action.

In Hong Kong land has great importance as a state owned and controlled resource, and specifically as a public-sector asset that monopolizes it. Yet private redevelopment is favored by the rather authoritarian public sector because it generates significant state revenue from the physical and economic intensification of sites. Gentrification is not on the public agenda but rather is a significant outcome of much-desired redevelopment activities. A relevant 'institution' of state-led gentrification includes the state monopoly regulation of the leasehold land management system. Higher development densities generate higher fiscal revenues from the same plot of land (but also higher revenues for developers); therefore, incentives to densify are ever-present. Although the Urban Renewal Agency is an entrepreneurial institution and executes redevelopment projects in partnership with powerful developers, significant aspects of its operations are designed to mitigate the financial and other distress of displaced

Figure 3.4: Redevelopment of Lee Tung Street in Hong Kong, 2011 (Photograph by Hyun Bang Shin)

lower-income owners and tenants, and to facilitate their remaining in their neighbourhoods.

In mainland China, a number of former rural villages have gone through some sort of informal densification, helped by the investment by original villagers who aim to extract rents from leasing part or whole of their densified properties to migrant tenants (see Tang and Chung 2002; Wang, Wang and Wu 2009). Very often, these urbanized former rural villages are stigmatized as undesirable, unhygienic or sources of urban crime, and become targeted by urban governments for redevelopment, especially when such places come in the way of city promotions such as mega-events (see Shin and Li 2013). They also occupy strategic locations as urbanization deepens, thus showing an increasing degree of potential profit when their lands are put into a 'higher and better' use, accommodating commercial, residential and business functions to cater for the needs of urban elites and more affluent households (Chung and Unger 2013). The redevelopment of such former rural villages may also be seen as an example that contradicts the experiences of the West in the original framing of rent gap expansion and closure, with disinvestment in the earlier phase and subsequent reinvestment in the final stage. The heavy household investment made by villagers goes against the notion of disinvestment as argued by most cases in the global North. It is the case of how 'obsolescence' and 'stigmatization' lay the foundation for such former rural villages to experience redevelopment in order for urban governments and developers to capture increased ground rents when under-utilized former village lands (from the perspective of those urban governments and developers) are put into a 'higher and best' use.

THE RESURGENCE OF RENT GAP THEORY

Rent gap theory was largely criticized during the 1980s and 1990s for its supposed rejection of individual and social agency as factors that determine neighbourhood redevelopment (Duncan and Ley 1982). Currently, we know that an enlarged rent gap is not a sufficient condition for gentrification to exist, but it explains important situations, the most important one being the fact that amid redevelopment processes, an important number of owner-residents will not receive enough compensation and will not be capable of affording decent relocation, while they are compelled to sell their landed properties (or cede their land use rights) cheaply to more powerful economic agents.

As we have shown in several cases above, and also in greater detail in Lees, Shin and López-Morales (2015), researchers in the global South visualize a gentrification that depends largely on state policies and huge economic investment in redevelopment. As such, the classic 'icon' of the middle-class gentrifier is weak if compared to the great land purchasing power of developers. The result is that the content of the terms, 'state-led' gentrification and 'gentrifier', need to be reconsidered in order to comprise a different kind of relation between the agent and the producer. In fact, Eric Clark's (2005) use of the word 'users' rather than 'residents' seems quite appropriate to this goal. We also agree with Slater (2015) who argues that it is futile to advance the complaint that the rent gap cannot tell us anything about the new middle classes that gentrify city space, as this theory was never designed to do so, and as we now know middle-class behaviour is not a sufficient precondition for gentrification everywhere (we tackle the 'middle-class' issue in Chapter 4). In fact, we think at some point it is impossible to deny the power exerted by bankers, developers and state officials both in the devaluation and revaluation process of urban change. As Slater (2015: 19) says,

> [t]he rent gap is fundamentally about class struggle, about the structural violence visited upon so many working class people in contexts these days that are usually described as 'regenerating' or 'revitalizing'... Without the rent gap, we would not understand this class struggle like we do, nor have such a clear set of critical analytic optics through which to interpret and challenge cycles of investment and disinvestment in cities.

Contra many other sophisticated post-structural theorizations, the rent gap is a simple idea that anti-gentrification activists can read, understand and debate, and then apply to their everyday realities for nurturing their struggle to challenge the dominant systems of urban segregation (see London Tenants Federation, Lees, Just Space and SNAG 2014):

> Identifying rent gaps, and identifying those institutions creating them with a view to capturing profits from them, is clearly vital to the formulation of strategies of resistance and revolt. Therefore it is a critically

important challenge for scholars and activists, together, to identify precisely where developers, owners and agents of capital and policy elites are stalking potential ground rent; to expose the ways in which profitable returns are justified among those actors and to the wider public; to raise legitimate and serious concerns about the fate of those not seen to be putting urban land to its 'highest and best use'; to point to the darkly troubling downsides of reinvestment in the name of 'economic growth' and 'job creation'; to examine the possibilities for concerted resistance; and to reinstate the use values (actual or potential) of the land, streets, buildings, homes, parks and centres that constitute an urban community. (Slater 2015: 20)

SPATIAL CAPITAL AND GENTRIFICATION

In the global South, public policies are more important causal factors of gentrification than they were in the global North in the 1980s, and there is little understanding of the public investments and changes at the macro scale that have increased accessibility and mobility into certain neighbourhoods, with an artificial generation of 'spatial capital' (Rerat and Lees 2011). What characterizes the gentrification of vast urban regions in cities of the global South is its widespread occurrence after the creation of new centralities in previously undervalued urban areas, but whose location has become strategic now for higher-status office or residential revitalization, or service provision. In newly developing economies and expanding urban markets, this dependence on infrastructures is due to the traditionally profound shortcomings of services, amenities, and access to transportation, in the formal or informal neighbourhoods that exist (and that can be peripherally or centrally located), and are now undergoing intensive redevelopment of their infrastructure.

Drawing on Urry (2007), Blanco et al. (2014) suggest that mobility is relative and in fact there are a range of possible mobilities, according to the place where people live and work, plus other contextual constraints like time or economic or cultural assets. People's access also depends on the means of transport and communication available for them that makes some areas more desirable than others in the city, hence more likely to be appropriated by a dominant class that holds higher cultural and economic means. Spatial capital is in fact an outcome of the interplay between access, competence and appropriation. Access is related to the range of possible mobilities according to place, time and other contextual constraints; competence refers to the skills of individuals and groups, while appropriation refers to the strategies, motivations, values and practices of individuals, including the way they act in terms of access and competences (be they perceived or real) and how they use their potential mobilities. It is curious that this theory of 'spatial capital', which was conceptualized by Rerat and Lees (2011) based on a well-developed country like Switzerland, has garnered so much interest in a developing country like Blanco's Argentina.

Policy-led changes in transport access allow one class to increase their chances of taking over a certain territory while the other class loses access to it. These changes can be seen as a politico-technical form of segregation too. And among the other things that the concept of spatial capital can inform us of is an understanding of what differentiates the gentrifiers from the gentrified is not only their economic power to purchase, assemble and/or speculate with land and properties, but also, crucially, their class differentiated material and immaterial accumulation of access to a wider range of spatial capital. In Rio de Janeiro, Cummings (2015) discusses the idea that the *Metrocables* in downtown Rio favelas, like Morro da Providencia, are elements that enclose spaces of informality and poverty within a forced socio-spatial formalization and commodification of the central space of the metropolis. In Santiago, Chile, the *Efecto Metro* has been a highly decisive policy for opening new spatial niches of operation to real estate developers and housing consumers, enlarging rent gaps, and enabling new social classes to capture spatial capital, so generating indirect forms of displacement (López-Morales 2010). In Manila, the Philippines, Choi (2014) observes how the urban poor located in informal settlements are exposed to the risk of displacement by public transportation projects, especially a regional train line that valorizes not just the land surrounding the infrastructure but also the locational opportunities generated (we discuss more of this in Chapter 7 on Mega-gentrification and displacement).

Centrality and spatial capital are important key issues in arguing against the notion that planetary suburbanization is becoming the leading force, while gentrification is becoming somehow increasingly irrelevant (Keil, 2013). However, as Andy Merrifield (2013a) argues, if we think of the rise of multiple centralities and the realignment of centre-periphery relationships, as Rerat and Lees' (2011) 'spatial capital' theory tells us to do, then we can also assume that new 'urban frontiers' are going to be everywhere where such new centralities emerge. This is where transport access and the creation of spatial capital rises as a leading variable in attracting capital (re)investment and producing contestation between existing inhabitants and gentrifying forces, particularly when the reinvestment cycle kicks in.

POST-CRISIS GENTRIFICATION

We claimed earlier that the subprime mortgage financial crisis, which exploded first in 2008 in the US and Western Europe and resulted in skyrocketing housing prices, sending millions of people into default, is a recent vivid example of capital switching. On the one hand, the increasing complexity and scale of the global industry of subprime debts, and on the other the way national states 'rescued' the private bank sector after the crisis, are both forms of the speculative creation of capital. In the US and Europe, while the price of urban properties soared by a hypertrophied demand fed by 'risky' credits, some members of the most vulnerable social groups were able to inhabit neighbourhoods and houses that were previously financially unaffordable to them. What seems to have changed in European countries like Spain and Greece and the US after the crisis, is that millions of households were displaced as the dwellings they had bought were repossessed by banks. But those same devalued emptied properties, are now being massively re-bought by *buitres* (vulture) international funds, due to the fact that large rent gaps have been created, and so those countries have started to see ascending land and property prices again. The housing situation in the developed countries of the world most affected by the 2008 crisis has considerably changed in the last ten years, and vast neighbourhoods are facing new forms of exclusion and certainly gentrification (Aalbers 2011).

Mueller (2014) discusses the gentrification of the vast, mostly poor and mostly black residential areas of Washington DC, where the liberal attitudes of the middle classes that raid into these neighbourhoods (which resemble Jane Jacobs' descriptions of bohemian barrios, and that have become a focus of desire for the liberal middle classes) are much less important than the deeply structural economic forces that sustain these changes. Washington DC epitomizes the US critical economy: 15 per cent of families earn US$200,000 or more a year, but 15 per cent lives under the poverty line. There, government-led gentrification has trickled down to small home buyers, as neighbourhood recovery has been a takeover planned by large business interests who fund projects with tax exemptions. Real-estate values have soared again, and speculative condo developments have begun to replace single-family homes, very much resembling what we see today in Santiago's or Mexico City's gentrifying inner quarters. In Washington DC, the average price of a residential property in 2000 was around US$150,000; in 2009, it was over US$400,000, and then prices increased again by over 10 per cent in 2013. Currently, the local professional, higher-income classes can secure mortgages in gentrifying neighbourhoods and buy property, then ride those property values to secure their position as middle class. The mostly black, lower-income strata cannot do this. As Mueller claims, this market is economically racist, as it was before the crisis, with poor black households living the illusion of becoming middle class. We can see here how gentrification is very much a top-down affair, not a 'spontaneous hipster influx orchestrated purely by the real estate developers and investors who pull the strings of city policy, with individual home-buyers deployed in mopping up operations' (Mueller 2014: web article). Public policies like the construction of a trolley line (spatial capital again) advertise neighbourhoods as up and coming, with the possibility of skyrocketing property values, rapidly creating a mix of property circulation.

As the latter case shows, financial capitalism recovers or takes over from crisis through 'dirigiste', entrepreneurial state roles in gentrification. This was also seen in the early 1990s in Russia once the introduction of a free-market economy took place (Badyina and Golubchikov 2005), where gentrification in central Moscow was (and still is) the product of a complex interplay between market pressure aiming to meet the demands of Moscow's new post-Soviet economy, the demand by new upper-middle classes and oligarchs, and Moscow government's entrepreneurial orientation. And it can also be seen in East Asia, where the case of Taipei's gentrification is a good example of this in the current era of political shifts, characterized by privatization, deregulation, marketization and individualization (Jou et al. 2014). Huang (2015) stresses how public-owned assets located in the city centre of Taipei have played a major role in gentrification, as direct government-led policies include the rent-seeking privatization of public land and buildings, previously aimed at public housing, also opening the door for the participation of private developers in inner areas' urban renewal and making public investments to increase the locational advantages of certain spaces. Taiwan's urban and housing policies have also comprised large doses of class-motivated violence or revanchism (Atkinson 2003) during the last quarter century, for the sake of finance and property capital reproduction (Jou et al. 2014). In 1998, the Taiwanese state started a trend of policy modifications towards the establishment of a new housing policy that, over the past decade, has turned centrally located public land and housing into enclaves for the super-rich and elite professionals, in many ways resembling a form of hyper-gentrification, which has also happened in London, New York, and San Francisco. *Hyper-gentrification* is an accelerated taking over of land which is bigger, faster, and much more destructive than the traditional narratives of gentrification. Due to this, a large number of public housing units have become upscale commodities on the real estate market, as the Taiwanese state has sold them to large-scale market agents, turning them into enclaves for the wealthy and elite professionals and also pushing up dwelling prices in surrounding areas (also see Lees 2014b on the same process re. London's public housing estates). Taipei has become increasingly inaccessible for middle-class or young people to live in. In 2008, the ratio of median housing prices to median family annual income in Taipei rose to ten, and in 2012, further rose to fourteen, being one of the highest in Asia.

CONCLUSIONS

Amid a harsh, global, financially created economic crisis, the theory of capital switching seems more relevant than ever as a measure for crisis resolution, in the process of capital accumulation. In this vein, the increasing scale of gentrification worldwide seems to be an effect of the current needs of financial real estate speculators. Some are still able to deal in 'subprime' risky assets, most are directing their attentions towards upscale redevelopment and consumers of considerably higher income, thus skyrocketing prices in certain neighbourhoods or cities.

Within a framework of international neoliberal policy prescription, capital switching accommodates various spatial investment strategies attuned with local-level public policies of land upzoning, social cleansing, real estate privatization, transport accessibility, and so on; in order to facilitate the revaluation of land and real estate, be this in newly post-socialist regimes, or already highly neoliberalized national economies. To some extent, global speculative gentrification has replaced subprime markets.

The current scales of gentrification range from micro to mega. Both scales lead to the re-writing of the urban landscape. The key question is who is making these changes, and at the cost and for the sake of whom these changes are being made. In fact, in the present global economic crisis, the devalorization of capital invested in specific parcels of urban land leads to a situation where the ground rent capitalized under current land uses is substantially lower than the ground rent that *could* potentially be capitalized if the land uses were to change. If no financial capital is available *in loco* (like in current Spain or Greece, or post-soviet Russia), global financial capital is readily available as it experiments by moving around the world. It is at this point that it operates at the state level, opening new spaces for financial reinvestment, destroying outmoded physical and social infrastructures, blighting or stigmatizing spaces, and upzoning places, etc. But not all the cases of gentrification can be effectively deemed 'neoliberal', for other forms of dirigiste-states can be found, especially in East Asia.

The 2008 global economic crisis has made governments set policies aimed at reducing financial housing affordability for lower-income segments, and increasing the power that large-scale economic agents deploy in the land and housing markets. We can see everywhere, from London to Rio de Janeiro, and from Taipei to Santiago, that when redevelopment and rehabilitation become profitable prospects again, capital begins to flow back into disinvested land, and then substantial profits can be absorbed, very often at the expense of low-income people currently occupying that land. Rent gap theory is currently proving useful again to measure the effects of these massive economic projects, especially addressing the unequal capacity urban agents have to capture the economic riches provided by land, and also addressing the way the essentially class-monopoly absorption of the ground rent leads to differential rates of direct and indirect displacement. As such, it seems always necessary to uncover how and where rent gaps are emerging in different societies across continents, and to test, extend, complicate, and challenge initial frameworks of this theory by taking it into, and comparing it across, new geographic, empirical, and analytic terrains. This chapter has aimed to do so. We think that much more research and reflection are needed on the connections between land, rent, displacement, and the everyday production of urban segregation, namely gentrification. We do not mean that private economic profit is the only or main causal factor of gentrification everywhere, but we do claim that the new economics of gentrification explain fundamental questions like the induced inequalities of land and housing affordability, the essentially neoliberal economic mechanisms of displacement, and the financialization of urban change. Indeed, we believe there is a pressing need for more theoretically guided comparison and dialogue across national borders to avoid getting locked into the parameters of local debates (this is one of the risks of a too empirically oriented 'comparative gesture', like Robinson's 2011a, see Lees in press), especially in order to understand the phenomenon of speculative landed developer interests as probably one of the main forces reshaping cities everywhere in the capitalist, and indeed non-capitalist, world.

4 Global Gentrifiers: Class, Capital, State

Class and the idea of one class (new middle-class gentrifiers) displacing another class (the working class) has long been central to Anglo-American debates and discussions about gentrification (Lees, Slater and Wyly 2008). If gentrification is now global, we must investigate whether it involves a (global) gentrifier class and if it does, how uneven its intervention is going to be across the globe (also see Bridge 2007). Class is one of the most contested concepts in social theory. In this chapter, we do not discuss the middle class per se but rather the 'new' middle class as global gentrifiers. Importantly the term 'new' middle class in the literature on gentrification from the global North refers to a post-industrial, postmodern, 'new' middle class whose values and lifestyles were very different to the traditional middle class (see Ley, 1996, for a detailed account). By way of contrast, the term 'new' middle class in the literature on gentrification from the global South refers to a newly emerging or expanding, modernizing middle class with new spending power and associated interest in consumerism. Europe and North America have had welfare systems in place for longer than the 'Third' World, as a consequence the middle class is better established there, and has reproduced and mutated (fragmented) over time. By way of contrast, in Asia and parts of Africa the emergence of a middle class is coincident with the post-1970s periods of rapid economic growth with much improved social access to state- and privately provided welfare. In the case of Latin America, as we argue below, the emergence of the middle class is the effect of the rapid economic growth that began in the early twentieth century.

A 'new' middle class (like in the Anglo-American literature on gentrification) is rarely compared to a traditional middle class in cities of the global South. Indeed, in many cases a traditional middle class did not exist as such: in other places where one did exist, the old and new middle classes are blurred. The politics and lifestyles of 'new' middle-class gentrifiers in the global South are therefore not necessarily in reaction to the politics and lifestyles of a pre-existing (conservative) middle class (see Ley 1996; Butler 1997). What is interesting, however, is that the lifestyles and in many cases the (now rather suburban) values of contemporary gentrifiers in the global North and South are becoming more and more alike. As such, can we identify an emergent global gentrifier 'new' middle class? Has gentrification become generalized enough so that gentrifiers have gentrified minds? How important are they as individuals, as human agents, in the process of gentrification around the globe today? And most importantly, has state-led gentrification overtaken the importance of individual gentrifiers?

Some are now arguing that for the first time in history, a truly global new middle class is emerging. Homi Kharas, a senior fellow at the Brookings Institute and a former World Bank economist, and Geoffrey Goertz (2010) project that the new global middle class will rise from 2 billion today to 5 billion by 2030. Until recently the world's middle classes have been located in Europe, North America and Japan, but that is changing rapidly. The European and American middle classes are predicted to shrink from 50 per cent of the total to just 22 per cent. Rapid growth in China, India, Indonesia, Vietnam, Thailand, and Malaysia will cause Asia's share of the new middle class to more than double from its current 30 per cent. By 2030, Asia will host 64 per cent of the global middle class and account for over 40 per cent of global middle-class consumption, mostly fuelled by growth in East Asia. BRIC countries have seen different patterns of economic growth which have brought different social changes. The emergence of a new middle class is more recent in some countries than others. Obviously, estimations of a new middle class are fraught with difficulties, especially when measuring the size of the middle class, let alone defining it, and this is subject to debate (see Chapters 1 and 2 in Chen 2013). Furthermore, such an estimation often reflects state aspirations, making middle-class expansion a political project. For instance, in mainland China, middle-class construction has been a national project that received great emphasis by the Party State (Tomba 2004; see also Zhang 2010). As such, '[t]he middle class that the Chinese Party State envisages is clearly the most affluent in China's urbanising society, whose lives are detached from the masses' (Shin 2014a: 513–14).

In Latin America, especially in Brazil, Chile, Argentina, Uruguay and Mexico, a middle class was formed between the 1930s and 1940s by developmentalist states and massive access to public university education, so there the rise of a middle class is in fact not new. Indeed, in Brazil, by the early 1980s the middle class made up 15 per cent of the population, that figure has now increased to a third. Countries like Argentina and Mexico also grew their middle-class populations in the 1970s and 1980s. In East Asia, the middle classes also grew post-war: in South Korea, the proportion of professional, managerial, and clerical workers (not including sales employees) increased from 6.7 per cent to 16.6 per cent between 1963 to 1983; in Taiwan, the same workers increased from 11.0 per cent to 20.1 per cent between 1963 and 1982; and in Hong Kong, white-collar workers increased from 14 per cent in 1961 to 21 per cent in 1981 (Koo 1991). The rise of the middle classes was thus a

> significant element in the evolution of the new industrial order in the East Asian countries. Concentrated in large urban centers, they shape the dominant pattern of consumption and urban life styles; they make new political demands; and they spawn new political debates and

tactics. The political stance of the middle classes has been a critical element in the recent transitions from authoritarian rule in South Korea and Taiwan. (Koo 1991: 485)

So a new middle class making new political demands and with somewhat similar values to those that David Ley (1996) talked about as 'gentrifiers' in Canada was evident in East Asia at the same time, *not* later. Growth in China, however, lagged behind but is accelerating. In 2009 only 10 per cent of the population was middle class, in 2014 it was 43 per cent, and by 2030 it is predicted to soar to 76 per cent (see Tomba 2004; Chen 2013; and Shin 2014a on the middle class as a state project to expand the consumption base). Middle-class households are considered to be more urbanized than poorer households, and in mainland China, urbanization, middle class-ization and the proletarianization of rural farmers go hand in hand (Shin 2015).

As we entered the twenty-first century, Rofe (2003) argued for an understanding of gentrification as a strategy of social capital accumulation based upon an appeal to all things global and cosmopolitan, he discussed the emergence of a gentrifying global elite. For Rofe (see also Butler 2007b), gentrifiers reacted to global processes by creating identity in place. These *agents* of gentrification performed habitus in a time of heightened capitalism and spatial transformation in social, political and economic structures. These place-based performances of identity have been seen as increasingly 'global' in character. Indeed, Butler (2010: 4) has argued for a closer and more focused 'understanding of what it is that particular groups feel has changed for them with the onset of global neo-liberalism, social and economic uncertainty and a general detachment from the norms and values with which they were raised and may now be of little guidance in raising their own children (Sennett 1998)'. His question, however, is a very Euro-American one, harkening back to some sense of social and economic stability that was, and certainly is, not necessarily the case. The 'new' middle class in rapidly modernizing countries are more likely looking forwards, not backwards. Many of these are not the gentrifying global or transnational elite that Rofe (2003) discussed. Like the traditional Western middle classes, their class position, however, is unstable and chained to enormous amounts of debt as the cost of living escalates in modernizing societies.

Indeed, Bridge (2007) argues that there is no such thing as a 'global gentrifier' and Davidson (2007) argues that gentrification as a global habitat is less about class formation and more a corporate creation. Davidson (2007: 491) argues that global gentrification is 'less generated by the agency-led creation of habitus (Butler 2003, 2007b; Rofe 2003) . . . and more by the commodified production of habitat. Furthermore, he contends that gentrification globally is 'less a sociological coping strategy or form of class reproduction', but rather it is 'a capital-led colonialization of urban space with relations to globalization in terms of architectural design, investment strategies, social-cache-boosting marketing strategies and "non-local global" lifestyles' (Davidson 2007: 493). Global commodified forms of gentrification are capital-led developments of urban space that do not need to perform discursive practices (Bridge 2001) nor deploy social capital (Butler 2003). It is corporate property developers who have utilized globalization via architectural aesthetics and place marketing to extend the scale and scope of gentrification.

We agree with Davidson (2007) and argue that planetary gentrification is a capital-led colonization of urban space related to globalization and neoliberalization. From this stance, middle-class gentrifiers can be consumers and sometimes crucially deploy their class power to transform neighbourhoods, but they do not always do so and are not always a necessary condition for gentrification to happen. Then, it is our contention in this book that the role of the middle classes in gentrification needs to be reconsidered. But, gentrification is not simply a corporate creation either; it is more often a creation of the state – local and national. The state often does the preparation work for corporate capital to follow, similar to classical gentrification in the global North where pioneer gentrifiers risked and tamed the frontier for corporate capital to follow. The scale can still be neighbourhood, but equally it can be an entire city. The emerged and emerging new middle classes still have a role to play, but this is more now as consumers rather than producers of gentrification.

POLITICS

It has long been claimed in the global North that new middle-class gentrifiers were left liberal and held specific, often counter-cultural, beliefs and values that desired political and social reform (Caulfield 1994; Ley 1996; Butler 1997). In the global North, the social contract of the liberal state was with the middle class, not the rich or the poor, but this is breaking down in a privatizing, neoliberalizing world. The Occupy movements that Merrifield (2013b) discusses were/are very much about the breakdown of the social contract with the state. Social, economic and political changes have undermined the connection between the state and the middle classes in the global North, but what about in the global South? We cannot make assumptions that the planet's new middle classes will behave like the previous new middle classes in the global North. The assumption is that as people become more middle class they become more democratic and push a more liberal agenda, but in China it is suggested that the middle class there accept a more autocratic regime in return for the stability of their middle-class lifestyle. By way of contrast, the new middle classes in countries like South Korea have demonstrated liberal values. They 'have acted as a progressive democratic element in political transitions but the specific meanings and goals of democracy they projected differed significantly from the main concerns of the working class' (Koo 1991: 486). The temporal emergence of these groups under different economic and political systems is key to understanding: '[n]urtured by the state and being the major beneficiaries of the state-led urban accumulation and economic development, China's middle-class populace is unlikely to be an agent of social change; for as long as the state protects their wealth and ensures their current economic position, they would be unlikely to join up with the rest of the society in what Andy Merrifield (2011) refers to as "crowd politics"' (Shin 2014a: 514). Of course, the rise of any middle class brings about new notions of rights awareness. The new middle class makes new political demands to expand and protect their own rights, but particularly in the context of economic development led by strong, authoritarian states, there is always the trade-off between individual freedom and economic affluence. How the middle classes respond to state elites depends on what kind of pacts are established between the ruling elites and the middle classes (Slater, D. 2010). In Singapore, the People's Action Party has been ruling the country for decades by nurturing the middle class and meeting their material aspirations, effectively creating what Rodan (1992) refers to as 'dictatorship of the middle class'.

It is worth considering the seemingly similar, and also different, political agendas of new middle-class gentrifiers in the global North and South. One that featured strongly in the gentrification of Vancouver in Canada, and has been seen to be associated with pioneer gentrifiers in other Northern cities, is the 'livable city' agenda (see Ley 1987, 1994, 1996; also Lees and Ley 2008 and Ley 2011). Interestingly this same 'livable city' agenda (even if it is not called that) is being played out in the global South but differently from the way it did in the global North.

So how do 'livable city' agendas in gentrifying cities in the global South compare? Anthropologist Amita Bavisker (2007) discusses the remaking of Delhi through a series of judicial orders: the Supreme Court of India has initiated the closure of all polluting and non-conforming industries in Delhi, displacing an estimated two million people in the process. At the same time, Delhi's High Court ordered the removal and relocation of all squatter settlements on public lands, displacing more than three million people in a city of twelve million people. These actions were set in motion by the filing of public interest litigation by environmentalists and consumer rights groups, who have emerged as an organized force in Delhi, in relation to issues such as urban aesthetics, leisure, safety and health. For the bourgeois environmentalist, the new, modern, Delhi must move production activities outside of the city, like many post-industrial cities did in the global North. City spaces are to be reserved for white-collar production and commerce, and consumption activities. But unlike in Vancouver where the livable city was a quality of life/green environmentalism, in Delhi it is a consumption-oriented environmentalism. Bavisker's (2007) discussion of 'bourgeois environmentalism' points to the detrimental impacts of this agenda on the poor, in terms of their displacement. Here, the 'livable city' that bourgeois environmentalists in Delhi push for, has been inculcated into a neoliberal programme of gentrification. Clean air and clean spaces are more important than shelter and employment for the poor.

Ghertner (2011), however, disputes the idea that India's new middle class mobilized to reclaim urban space from the poor, gentrifying and 'cleaning up' its cities. Using the example of Delhi's Bhagidari scheme, a governance experiment launched in 2000, he argues that urban middle-class power did not emerge from internal changes within the new middle class itself, but was produced by the machinations of the local state. He shows how Bhagidari realigned the channels by which citizens access the state on the basis of property ownership, thus undermining the electoral process dominated by the poor, and in so doing privileging property owners' demands for a 'world-class' urban future. He argues that Bhagidari 'gentrified' the channels of political participation, respatializing the state by breaking the informal ties binding the unpropertied poor to the local state and thereby removing the obstacles to large-scale slum demolitions. Here, the state itself was heavily implicated in creating and gentrifying political participation, rather than the middle classes themselves being seen as agents of gentrification. Elsewhere Bose (2013, 2014) talks about the neoliberal vision of Marxist leaders in Kolkata/Calcutta as aligned with the bourgeois desires of the middle class (see also Kaviraj 1997).

Lemanski and Lama-Rewal (2013) have critiqued the dominant image of India's middle class as 'exemplified by glitzy shopping malls and international travel' as conflating India's new and tiny globalized elite with a large and heterogeneous middle class. Indeed what and who constitutes India's middle class remains uncertain and different for different authors. Authors like Fernandes (2006) who have focused on the 'new middle classes' (or what Brosius (2010: 1) calls the 'arrived middle classes') ask for dedifferentiation. Importantly, Lemanski and Lama-Rewal (2013: 96) point out that the Indian new middle class are 'a notion encompassing a normative vision of Indian's future society' as educated, upwardly mobile with Western consumption habits, if not Westernized values. Fernandes (2006) calls them 'consumer-citizens'. But Lemanski and Lama-Rewal (2013) argue that this really only represents the upper sections of the middle class. There remains a lot of ambiguity which demands further deconstruction of the Indian middle classes and their political agency.

Koo (1991) discusses both the similarities and differences between South Korea's new middle classes and those in the West. He points to the fact that there are also two diverging views of the politics of the new middle class in South Korea – some argue they have been politically very progressive and acted as an important democratic force, while others argue that they were basically conservative and maintained authoritarian regimes. In terms of the former, he talks about the fact that the South Korean new middle classes played a much more significant role than their counterparts in the West did in working-class struggles. The reason, he argues, is historical, for the Korean new middle class emerged as a significant social stratum before the bourgeoisie established its ideological hegemony and before industrial workers developed into an organized class ... Occupying a relatively more autonomous position from state control than wage workers do, intellectuals, students, and other white-collar workers participated actively in political struggles against the authoritarian regimes, exerting strong influence on the consciousness and organization of industrial workers. (pp. 486–7)

Koo (1991) is a wonderful early comparative urbanism of the new middle class in South Korea and the West. The rise of the developmental state, borrowing technology from abroad meant that class positions and relations were somewhat different. He makes some important comparisons: timing – in the West the new middle class emerged *after* the bourgeoisie had established its hegemony and the formation of a large and very political working class, largely due to the benefits of the welfare state. In the newly industrialized countries of East Asia, the new middle class emerged at the same time as, and in some cases before, the rise of the capitalist and the working classes. Due to their high status, a relic of the Confucian tradition, Korean intellectuals (with social status and moral superiority above other groups) filled a gap in terms of ideological formulation and social movements. It was Korean intellectuals who became increasingly politicized and criticized the authoritarian political structure (see Figure 4.1). The counter-cultural political activities of South Korea's new middle class certainly echoed those in North America:

In April 1960, the middle classes, both white-collar workers and small business owners, provided strong support to students, who eventually succeeded in toppling the dictatorial government of Syngman Rhee. In the 1971 presidential election, a large majority of the urban middle-classes cast votes for the opposition candidate, Kim Dae-Jung as a protest against Park Chung Hee's attempt to remain in power

Figure 4.1: Student protests in Kwangju, South Korea, 1992 (Photograph courtesy of Namkug Kim)

indefinitely. The middle classes revolted again in 1985... In this election, the New Democratic Party, which had been formed for no more than three months by old opposition politicians whose civil rights had just been restored after several years' incapacitations, swept across urban middle-class districts, although it also drew a strong support from the working class. The most dramatic incident of the middle-class involvement in the Korean democratic struggles occurred in June 1987, when a large number of white-collar workers joined the students' street protests against Chun's refusal to amend the constitution for a direct presidential election. (Koo 1991: 490–1)

Protests on the streets were a large part of the counter-cultural politics of the 1960s that David Ley (1996) and others discussed as conditions of gentrification in the West. Koo (1991) argues that the diverging theses on the new middle class in South Korea, as to whether supporting democracy or the status quo/authoritarian regimes, mirrors debates about the political character of the new middle class in Europe and North America, that middle-class politics are not stable, nor consistent, that they have contradictory material interests, and so on. Marxists would argue that this is because the middle class is a class in-between the bourgeoisie and proletariat, their politics oscillating between the ideologies of the two. As Poulantzas has argued (1975: 289) 'the petty-bourgeois ideological sub-ensemble is a terrain of struggle and a particular battlefield between bourgeois and working-class ideology'. Koo (1991) explains the difference in terms of the fact that the middle class is not a homogenous social group. Lett (1998) argues that above all South Korea's urban middle classes (pre Asian crisis) sought improved social and economic status, a relic of traditional ways, making them quite different to the classic left liberal gentrifiers in London and New York City. In some ways, if one reviews Koo's timeline, it looks like South Korea's new middle-class transition from progressive to more conservative, in some ways mirrors David Ley's (1996) thesis on hippie to yuppie in Canada. This symbolizes the 'gentrification of the mind' (Schulman 2012), as a rebellious culture gets replaced with a conservative culture linked to mainstream consumerism. During the post-2008 financial crisis, the South Korean middle classes felt threatened in terms of maintaining their material wealth, and the middle class itself began to disintegrate, mostly downgrading due to the loss of their wealth (due to labour market restructuring and economic difficulties). As a result they have become increasingly conservative, supporting a conservative government that rhetorically keeps promising that the state will boost the property market – hence their political characteristics are quite different from what they used to be in the 1970s and 1980s. A good comparison would be Greece after its economic collapse and enforced austerity measures, where the middle class is also dissipating rapidly due to economic restructuring.

Our hypothesis is that middle-class politics in the case of East Asia, for instance, can be entwined with material aspiration and the speculative urbanization that builds on an increase in property wealth (also see Ley and Teo 2014, on the 'culture of property' in Hong Kong and Hsu and Hsu 2013 on the 'political culture of property' in Taiwan). We suggest that there is a close nexus between (a) political orientation in the context of condensed urbanization and industrialization (where ruling hegemony is in the making at the same time as middle-class construction); and (b) the material gains from gentrification, which lead to the affluence of the middle class, who invest in property assets as speculative urbanization takes place (see Shin and Kim 2015).

LIFESTYLE AND CONSUMPTION

Consumption-side explanations found in gentrification theory from the global North have stressed the role of choice, consumption, culture and consumer demand (see Chapter 3 in Lees, Slater and Wyly 2008 for detail). The key player in these explanations was/still is the 'new' middle-class 'gentrifier' and their consumption behaviours which are seen to trigger the gentrification process. David Ley (1996) argued that this 'new cultural class' was a product of the emergence of a new post-industrial society, and that they were liberals who sought to escape the banal routines, conservatism, and oppressive homogeneity of suburban life, seeking residence in the diverse inner city. Gentrification was the opposite of suburbanization. Lees (2000) termed this an 'emancipatory city thesis'; it has links to earlier notions of the European industrialized city as an emancipatory site where incomers could cast off the shackles of rural life (Lees 2004). Except for Caulfield (1994) and Ley (1996) gentrifiers cast off the shackles of suburban life. By way of contrast cities outside of the West, like those in Africa or Asia, were seen to be stuck in the past, in un-modernity, and as such, were not seen as emancipatory. Developmentalism played on their un-modernity.

When considering gentrification in the global South, this particular Western consumption thesis makes little sense (see Lees 2014a). For example, the large-scale suburbanization associated with the industrial/modern city in early to mid twentieth-century Euro-America did not happen in Chinese or Latin American cities. So, the idea of gentrification as a counter-cultural idea posited against the modern suburbs has no traction. Rather, the old German saying 'city air makes men free' (Stadtluft macht frei), free from the shackles of the rural, liberation from feudal master-servant relations, may have more purchase; or comparisons with the suburbs being built at the same time as inner cities being gentrified in cities in the global South, as in the case of mainland China, e.g. Beijing (see Fang and Zhang 2003) . With no experience of the post-war hegemony of suburban life, gentrifiers outside of Western cities will undoubtedly have different mindsets. In Latin America, urban development was disassociated from industrial growth and cities grew not through the densification of their city centres but through extension into peripheral areas. As such, the return of the middle class to Latin American inner cities and the subsequent displacement of the working class has been less common than in the US or Europe (Inzulza-Contardo 2012). In some countries, the new gender order of post-Fordism that pushed women into the workplace and created dual career households (see Rose 1984; Bondi 1991; Warde 1991) did not happen in tandem with the emergence of gentrification. For example, in China, gender relations have been fairly static since 1949 when there was a huge mobilization of women into paid work. The drive of socialism from the 1950s set up communal eating places, laundries and creches, meaning that women could work and manage family life more easily (Rai, 1995). By way of contrast, as the economy marketized, it became more difficult for women who could be pushed out of jobs in favour of men and profitability, and women's rights have long been subdued in the interests of class politics (see also Yang 2010).

Ren (2013) describes the new urban middle class in China, for example Ms. Zhang, who was 'young, educated, English speaking, career-orientated, technology savvy (and lived) a lifestyle not that dissimilar to (her) counterparts in other global cities . . .' (p. 172–3). Ren (2013) does not discuss the new urban middle class's ideological disposition to the gentrified inner city as anti-suburban; rather, it seems the suburbs just do not figure in the mindsets of these gentrifiers. Their lifestyle demands are much more orientated towards conspicuous consumption. As Wang and Lau (2009) point out, the professional middle classes in China are attracted by the image of elite life and thus willing to pay for the symbolic value of those elite enclaves which they can afford. They also possess cultural competence to decode and appreciate an urbane lifestyle.

Inzulza-Contardo (2012) concludes from his research in the historic Bellavista neighbourhood of Santiago de Chile that gentrifiers there are different to those in the global North – they are tenants rather than owners and they do not align with the white collar profile of 'Northern gentrifiers' in a high-income bracket. Rather they are light blue collar workers living in flats. The Chilean housing market advertises these flats as lofts, studio units and with luxury facilities like in the North. These gentrifiers are in tertiary sector jobs with new middle-low incomes. Their consumption is encouraged by globalized commodities. This is quite different to the more individualized, conspicuous thrift of classical gentrifiers in the global North (see Ley 2003). Anyway, in Chile, these pockets of historic gentrification are insignificant in comparison to the large-scale redevelopment of central areas of Santiago (see López-Morales 2010).

As we say in Chapter 2, the scale of gentrification in the South needs to be carefully taken into account as it does not replicate the classical micro-neighbourhood narratives of the North. For instance, Contreras (2011) claims that in Chile, the middle classes that 'consume' residential gentrification in the inner areas of the city respond to a wholly significant demographic and cultural change in Chilean society, with current household structures increasingly differentiated and responding to complex residential choices. Hence, Santiago's inner area now attracts younger households often with one or two children, young professionals who recently left their parents' nest (often first university generations), single or separated women with or without children, sexual minorities, and even relatively wealthy Latin American immigrants, among other groups. Their residential choices are based on the advantages of spatial proximity to jobs, services and public goods that exist in the central areas, and also more proximity to their social networks and family (also see spatial capital, Rerat and Lees 2011). Contreras (2011) argues that in Santiago there has been a succession of middle/professional classes, as residential gentrification is mixed with the studentification (Smith 2005) of certain areas that has been pushed by educational/real estate economic conglomerates.

Gay and lesbian gentrifications have been part and parcel of discussions of gentrification in Anglo-American cities, and indeed of gentrification as a counter-cultural, emancipatory practice (see Lees, Slater and Wyly 2008: 209–13). There has been much less discussion elsewhere on the globe, but it is apparent that gentrification and gay rights do have some purchase in some cities. Buenos Aires is one example where, as part and parcel of its touristification (see Herzer et al. 2015), gay guest houses (e.g. Lugar Gay in San Telmo) have opened where only gay men can book a stay. There is some interesting future work to be done on sexual politics and gentrification, especially in Latin America where same-sex civil unions have been legalized in Argentina and in some states in Mexico and Brazil. In Colombia same-sex couples now have inheritance rights and can add their partners to health insurance plans. Coming out of very Catholic countries, this is striking. Ironically these Latin American countries now seem ahead of the progressive curve in comparison to some Western

European countries. Contemporary Argentina is very interesting in terms of some of the parallels (and differences) between the sexual revolution and counter cultural politics in the global North, as Argentinian youth now practice sexual and other freedoms. The sexual politics of gentrification in the global South is food for thought. In East Asia, any relationship between gentrification and sexuality is mitigated by the state. Although gay sex is illegal in Singapore, it has been legal in China since 1997. As such in Beijing, for example, it may be that gay Chinese couples do not feel the need to congregate in a neighbourhood like the Castro in San Francisco. Finally, the bulk of the Anglo-American literature on gay gentrification, to date, has viewed it in an emancipatory light, but might it be time now to turn this on its head and consider whether gay gentrification leads to homophobia? In the same way that the social mixing caused by gentrification and mixed communities policy causes social tectonics, conflict and segregation might it be that gentrified gay enclaves increase external hostility?

Fernandes (2009: 220) offers some food for thought when thinking about the consumption thesis and gentrification in the global South:

> Existing studies of middle class consumption in India have tended to reinforce a consumer preference model of analysis – one that has addressed the relationship between economic change and consumer aspirations and behaviour. While such research has yielded valuable insight into questions of subjectivity and cultural change, there is a greater need for an understanding of the systemic relationship between consumption and the restructuring of state developmental regimes under liberalization.

In gentrification around the world today, the state plays a key role. When the (now) classic consumption theses emerged in gentrification studies in the global North (see Lees, Slater and Wyly 2010: 127–88), they were concerned less with the state and more with the impact of the transition to a post-industrial society and postmodern culture on the individualized consumption of a new class. The relative omission of the state and the focus on individualization does not aid conceptualization of the relationship between lifestyle and consumption and the state today – the role of the state is much more significant; post industrialism has progressed globally and what was individualized consumption (class gentrification) is now a global product (and indeed any individuality has now been mass produced). Gentrification researchers need to refine or look for a different conceptual model when looking at consumption and lifestyle with respect to gentrification in the global South.

HISTORIC PRESERVATION: GENTRIFIER OR STATE-LED?

> From Xintiandi in Shanghai, Asakusa in Tokyo, to Bercy in Paris, numerous historic districts are undergoing a process of Disneyfication, in which they are remolded into theme parks in the service of the global consumerist elite. (Zhang 2013: 156).

The relationship between gentrification and historic or urban preservation is no longer the now classic Jane Jacobs' (1961) social movement type story. Rather, it is now state-led. The classic story of historic preservation is of Jane Jacobs' fight to 'save' Greenwich Village in New York City from the federal bulldozers of Robert Moses' urban renewal. Jacobs' thesis on old neighbourhoods underlined a social movement of historic preservationism in cities across the US. She asserted: 'New ideas must use old buildings.' Osman (2011: 83) has described the desires of Brooklyn's 'brownstoners' (pioneer or first wave gentrifiers) as 'a new romantic urban ideal'.

The relationship between historic preservation and gentrification, however, was not just apparent in New York City (or the global North) but also in the global South. The earliest waves of gentrification in Istanbul in the 1980s, for example, demonstrated very similar processes. The early actors in the preservation and gentrification of the late nineteenth and early twentieth century two and three storey row houses in the inner city neighbourhoods of Kuzguncuk, Arnavutkoy and Ortakoy near the Bosphorus, were individuals and small investors. In the 1990s gentrification in Istanbul spread into Beyoglu, the historic centre of the city which held late nineteenth and early twentieth century row apartment buildings. Many of these neighbourhoods are now historic preservation districts. But like in the global North, the early actors (pioneer gentrifiers) were to be replaced by the state after 2005.

In Latin America, historic preservation and gentrification is also related. In fact, international forces enticed Latin American governments to engage in heritage tourism to repopulate and redevelop central cities as early as the 1970s. In 1977, the Organization of American States sponsored a conference that issued the Quito Letter, an agenda to transform historic centres into heritage tourism. Urban competiveness promoted by the Inter-American Development Bank urged local governments to undertake urban renewal. There were projects in Sao Paulo, Buenos Aires, Mexico City, Puebla, Santiago de Chile, Lima, Bogota, and more (see for example, Nobre 2003; Herzer 2008; López-Morales 2010). The International Development Bank, the Inter-American Development Bank, and the World Bank all targeted UNESCO-designated sites offering packages to local (and national) regimes for historic preservation. The regimes looked to the past, 'depicting the elites formerly inhabiting central areas as paradigms of civility', as such renovation was a 'class reinstatement through the "rescue" of ... elite values and virtues' (Betancur 2014). In Sao Paulo, Mexico City and Buenos Aires, in particular, extreme efforts at class replacement were made by condemning and acquiring buildings, plazas and monuments and renovating them for museums, cultural institutions and tourism. Governments agreed to coordinate the displacement of lower class residents as part of the requirements for the above banks lending them money. It was thought that this cleansing would attract the private sector into these city centres, but it has failed for the most part as it is not self-sustaining.

Significantly, where the Quito Letter was signed and where heritage conservation policies have been paramount, is the place where class-motivated revanchism has been better articulated outside the global North (see Swanson 2007, 2010). Inspired by zero-tolerance policies (see Chapter 5), Ecuadorean state-entrepreneurial urban regeneration projects have aimed to cleanse the streets of Quito (and also Guayaquil, the second largest city in Ecuador) of informal workers, beggars, and street children. Swanson shows that race and ethnicity are two devices through which the dominant middle-class oriented, globalized aesthetic for historical centres in the Andes unfolds, also showing how a project of 'blanqueamiento' or 'whitening' takes place, displacing already marginalized individuals from public spaces like streets and parks, and pushing them into more difficult circumstances outside of where informal jobs on inner city streets and mixed sources of small income exist.

Based on an analysis of Panama City's Belle Époque style Casco Antiguo, Sigler and Wachsmuth (2015) claim that there are novel pathways for gentrification there, much more connected not just with the in and out flows of financial capital but also the increasing real estate and cultural interest by a transnational middle class. As Panama shows, redevelopment capital is often connected to local housing demand not within a single city-region but along transnational pathways, thus creating housing reinvestment profit and also threats of displacement for the original population too. This 'globalized' market condition started to exist after a series of preservationist legislations were passed in the 1970s, and largely intensified with the 1997

UNESCO heritage designation that limited the potential for large-scale redevelopment, crucially catalyzing both international and domestic interest in the neighbourhood. Currently, in Casco Antiguo, rents in refurbished units range from approximately US$1,000 to US$4,000 per month, in a city where 57 per cent of residents pay less than US$200 per month in mortgages or rent. In addition, renovated apartments within the area are sold for between approximately $150,000 and $2 million. This is clearly beyond the capacity of most local inhabitants, even large numbers of the wealthy middle classes that anyway prefer newer suburban areas with a different set of amenities and cultural associations. The process of residential conversion and the developments in Panama Casco Antiguo are aimed at, and cater for, international tourists and expatriate residents. The neighbourhood's demographic composition has changed radically.

In some parts of the global South, gentrification has seemingly progressed the opposite way round from gentrification in the global North – progressing from large scale and new to small scale and old. As Lees (2014a) discusses – in China, for example, the 'new' middle classes are looking back to traditional architecture – expressing a yearning for traditional Chinese culture as a means through which to express their 'cultural taste'. Gentrification processes in China began as large-scale urban renewal, imitating Western modern, new-build, high-rise architecture, but have moved on somewhat and now show an interest in social responsibility through environmentally sustainable design and technologies, and in traditional architecture as seen in the preservation of 'hutongs' and 'lilongs' in inner city Beijing and Shanghai respectively. A hutong is a narrow street that has small single-storey houses coming off it, and the houses are normally made up of four buildings facing into a central courtyard. A lilong is a traditional urban alley community; the community is tightly interlinked – not just physically but also socially – because the residents also run the local shops and restaurants in the street. In the first wave of gentrification, many hutongs were knocked down to make way for new, dense, Western style housing developments; now they are more likely to be gentrified by rehabilitation/preservation rather than demolition and reused as new trendy cafes and shops, their market – the well-off young Chinese who want to feel cool (see Shin 2010, on Nanluoguxiang in Beijing). Often, historic neighbourhoods are preserved not through in situ upgrading but through wholesale clearance and reconstruction to mimic old styles, a process that locals sometimes refer to as 'fake-over', as happened in Qianmen, Beijing's centuries-old market place south of the Forbidden City.

Significantly, historic preservation has not been initiated by pioneer gentrifiers like it was in the global North (see Lees, Slater and Wyly 2008). Rather it is state-or developer-led. It is mainly historical buildings from the city's colonial pasts that are recycled in global city building processes. Ren (2008) discusses that from the end of the 1990s, Shanghai's city government began to support historic preservation by passing a series of preservation laws and turning large numbers of historic buildings into landmarks:

> the buildings listed for preservation were predominantly Western-style architecture from the colonial period before 1949. Buildings from the following socialist era are mostly regarded as worthless for preservation and eligible for demolition as they age. Only recently, responding to the criticisms about the lack of preservation for architecture from the socialist era, a few buildings built after 1949 were added to the preservation list. (p. 30)

The distinct temporal waves of Northern gentrification are all happening at once in China, and in a back to front way (not historic preservation to state-led new build, but state-led new build to historic preservation), underlain by the growth of a new middle class and a new consumerism. Ren (2008) makes clear that historic preservation in

Figure 4.2: 'Fake-over' of Qianmen, Beijing, 2014 (Photograph by Hyun Bang Shin)

Shanghai is very pragmatic; it is about potential economic return not only in terms of the local economic base and attracting investment and tourists but also from property appreciation. As she says: 'old historical buildings in Shanghai, once regarded as worthless in the frenzied development boom of the 1990s, have been rediscovered for their economic value' (p. 31). She talks about the Mayor of Shanghai's new slogan in 2004, which reads: 'Building new is development, preserving old is also development'. Historic preservation is therefore another and a more sophisticated tool than demolition employed by the government to achieve urban economic growth. It is not about cultural heritage so much as economic return.

Ren discusses the preservation of Xintiandi, two blocks of old Shikumen houses from the 1930s, which is regarded as the turning point from demolition (Chai) to preservation (Bao) in Shanghai. Unlike in early examples of historic preservation/conservation in London and New York City which saw pioneer gentrifiers investing sweat equity, international architectural firms played key roles, including one involved in the redevelopment of Faneuil Hall in Boston. Also, unlike in the examples in the gentrification literature from the global North, the preservation was not undertaken so people could live in the buildings: rather, they were redeveloped as shops, restaurants and night clubs. The only residential traces were in one of the houses which was converted into a museum, featuring an exhibition of everyday residential life in a Shikumen house in colonial Shanghai. Unlike in New York City, for example, where historic preservation was a bottom up social movement, in Shanghai, it is a top down instrument that sells a globalized image of modern Shanghai in the 1930s as one where wealthy middle-class families sent their children to English speaking schools, watched Hollywood movies and listened to Jazz (see Ren 2008: 36). The idea is to show that Shanghai was an international metropolis in the past and to feed this through to its global city future. The image is a cosmopolitan one, a hybrid of old and new. Historic preservation in Chinese gentrification is about creating a modern space, quite different then to that in the global North. Those displaced by the process, however, tell familiar stories to those displaced by gentrification in Western cities: 'Resident A (a man in his 50s) My family has lived here for three generations. Before the Liberation (1949), my grandfather bought the house. I have the contract. I don't want to live in the suburbs. There are no hospitals. It takes hours to get to the city and see a doctor by bus' (cited in Ren, 2008: 39).

Zhang (2013) provides insight into the comparative politics and urbanisms of historic preservation in Beijing, Chicago and Paris, demonstrating well the importance of specific structures of urban governance and the very different histories of urban transformations in cities. In Chicago, historic preservation was/is about improving the local tax base, in Paris it was/is about protecting French cultural integrity and national pride, whereas in Beijing it was/is a tool for local government to promote economic growth. In Beijing, from the early 1990s, when Beijing's Municipal government launched its citywide housing renewal project, many old neighbourhoods and buildings were demolished – in the face of widespread criticism, the government introduced preservation laws, designated preservation districts and added to funds to renovate historic structures. But the pace of demolition did not slow down, and Zhang (2013) argues that the result was a merely 'symbolic' urban preservation that 'serves primarily as a tool to smooth the functioning of the growth machine and to create a better global image for the city' (p. 24) (for comparison see He 2012; and Shin 2014b on Guangzhou; see Figure 4.3).

Figure 4.3: Redevelopment in the historic centre of Guangzhou, 2010 (Photograph by Hyun Bang Shin)

Elsewhere in East Asia, in Hong Kong, Singapore and South Korea heritage conservation or attention to historic neighbourhoods is gaining real weight for a diverse set of reasons: (a) heritage buildings are gaining higher exchange value; (b) waves of new-build gentrification are rendering historic dwellings and neighbourhoods more scarce (scarcity); (c) authenticity (albeit manufactured to a large extent) is claiming monopoly rents; (d) nationalism, as in China, has become a state-promoted ideology, emphasizing heritage preservation (selectively) (see Broudehoux 2004); and (e) Singapore sees selective conservation as a tool for ethnic policy (see Kong and Yeoh 1994; Yeoh and Huang 1996; Chang and Teo 2009; Chang and Huang 2005). And in Hong Kong, heritage conservation has been staged as a fight against state-led urban development, for example, the Star Ferry pier struggles (see Hayllar 2010).

CONCLUSIONS

Sassen (2006) identifies a new global class whose spaces are in cities in both the global North and South – but only in parts of these cities. She argues that a new global class in Sao Paulo, Johannesburg and New York interfaces more with elites in other world cities than with the populations in their own cities and countries. Her new global class has more in common with the highly mobile, transnational class (see Butler and Lees 2006, for a critique). Gentrifiers around the globe include these folk as we see in Panama and other cases, but gentrifiers (the new middle classes) globally are a much more diverse entity in terms of income (some are very rich, some lower-middle class), politics (some are liberal, some are conservative even authoritarian), and lifestyles (some are highly consumer-orientated, others much less so). There is a truly elite gentrifying class in, for example, African cities, and a less wealthy body of gentrifiers in, for example, some Latin American cities, like Buenos Aires. This presents a rather muddled picture globally. But what is clear is that gentrifiers in the global South are not the same as the classic 'new' middle-class gentrifiers in Anglo-America. In the Communist Manifesto, Marx and Engels (1848/1967) wrote that the rise of the bourgeoisie 'compels all nations, on pain of extinction, to adopt the bourgeois mode of production; it compels them to introduce what it calls civilization into their midst, i.e., to become bourgeois themselves. In one word, it creates a world after its own image' (p. 84). Global gentrifiers are now both bourgeois and the aspiring bourgeois, and globally this is what they have most in common. There are other convergences too, for gentrification around the planet seems to have adopted a suburban mindset, a conservative 'gentrified mind'. Very often national and local-level policies of urban change meet their goal and aim to satisfy their urban preferences as consumers, but this does not mean always and everywhere the middle class decisively exert class power to transform the neighbourhood (Davidson 2007). It is the nexus between planetary gentrification and planetary suburbanization and urban and suburban mindsets that needs further research globally.

The fact is, however, that the global gentrifier as a (human, individual) agent of gentrification is insignificant now not just in the global North but in the global South too. The key actor in planetary gentrification is the state – neoliberal or authoritarian. The growth of a new middle class globally in terms of the rising demand for new housing coupled with growing interests in real estate investment in producing new 'gentrified' urban forms is important, but it is the state that is the key constituent in gentrification in the global South and East. Seen from the South, the rise of the new middle class and its higher consumption power may not be a precondition for gentrification to occur, but a necessary condition alongside other structural causes, or even, gentrification can be an effect of broad policies oriented to reconfigure the spaces of social reproduction of cities and countries.

In addition, there are new 'gentrifications' by the super-rich, not the super-gentrifications that Lees (2003) and Butler and Lees (2006) have discussed that are connected to the global finance industries, but hyper-gentrifications caused by super-rich elites from the global North and the global South investing across borders, e.g. mainland Chinese investing in London and Hong Kong (Hong Kong's property boom post-2008 was helped by an influx of mainland Chinese hedging their risks and diverting their assets away from mainland China), or middle-Eastern oil wealth or Russian elites/oligarchs investing in overseas property markets (London's inflated property market has been blamed on these overseas investors). In London, in neighbourhoods as diverse as Kensington and Chelsea, Notting Hill, Hampstead and St. John's Wood, the wealthy middle classes are being displaced by a new super rich. Moreover, developers are undertaking new-build gentrifications across large swathes of London in anticipation of this money. These transnational investors are important. Indeed, it is our contention that planetary gentrification is produced less by global gentrifiers (the global North and South's new middle classes) and more by (trans)national developers, financial capital and transnational institutions including international financial organizations like the IMF and the World Bank, all of which the state courts.

5   A Global Gentrification Blueprint?

In the twenty-first century, some gentrification scholars declared gentrification was a blueprint being mass-produced, mass-marketed, and mass-consumed around the world. As the urban-rural dichotomy broke down, as the world became increasingly urbanized and desirous of an urban(e) lifestyle, even Third World cities were gentrifying (Davidson and Lees, 2005: 1167). Other scholars were a little more cautious:

> [T]here is something there in most parts of the world which we would recognise as some form of the phenomenon that we call gentrification. However, it is not like a bottle of Coca Cola with a registered trademark, which can be hauled off the shelf in any global housing market. (Butler 2007a: 164).

Smith (2002) argued that gentrification had become a 'global urban strategy', and that it had evolved in the 1990s into an urban strategy for city governments allied with private capital in cities around the world (p. 440). Gentrification had become bound-up in a global circuit of urban policy transfer where the promises of inner city 'revitalization' and 'renaissance' attracted national, metropolitan and local governments to promote the middle class-ization/elitization of central cities.

In this chapter we investigate whether a global gentrification blueprint is evident around the globe, for as Robinson (2011b: 22) says:

> Much is at stake in how we characterize the spatiality of urban policy transfer and learning. It is important to question understandings of policy exchange and innovation that are the inheritance of a deeply divided urban studies shaped by colonial and developmentalist assumptions.

By blueprint we mean a model or plan of action for something to be copied. But is 'gentrification' being copied/adopted in policies and plans around the world? Referring to Robinson (2002), Roy (2009: 820) paraphrases Robinson (2003: 275) stating that 'the enduring divide between "First World" cities (read: global cities) that are seen as models, generating theory and policy, and "Third World" cities (read: mega-cities) that are seen as problems, requiring diagnosis and reform' is a 'regulating fiction' that needs to be overcome in its 'asymmetrical ignorance'.

There has long been an assumption in the gentrification literature, a diffusionist logic, that gentrification (as both a process and a blueprint) is cascading downwards and outwards from metropolises in the global North:

> [gentrification] appears to have migrated centrifugally from the metropolis of North America, Western Europe and Australasia ... at the same time as market reform, greater market permeability and population migration have promoted internal changes. (Atkinson and Bridge 2005: 2)

This diffusionist logic has continued in more recent work by, for example, Peck (2010) who talks about the mobility of a fast policy of 'creative gentrification' around the world. In South African literatures on gentrification, the diffusionist logic is especially apparent: Visser and Kotze (2008) argue that, in South Africa, urban policies for regeneration and renewal show 'the imprint of global economic forces and urban redevelopment thinking' (p. 2569), and Winkler (2009) approaches inner-city regeneration in Johannesburg as 'nothing more than an euphemism for gentrification' driven by a blind belief in new global policies or, as her study's title suggests, the prolongation of the global age of gentrification as it moves toward Southern cities. Her analysis presents the market-driven regeneration of Johannesburg's CBD as a fitting example of how gentrification has become a focal point for global urban policymakers.

By way of contrast, authors like Dutton (2005: 213; also Dutton 2003) have complained that this image of a downward trickle from saturated global cities implies a single mechanism for gentrification everywhere and ignores the role of local context in 'mediating gentrification processes'. More recently, Teppo and Millstein (2015: 419), writing about Cape Town, do a good job of looking at the 'mediating context', arguing that 'the public justifications of, and discussions and disputes about, these processes differ greatly from those employed in the global North'. They point to colonial and postcolonial inheritance, the social hierarchies of a settler society, ideas of 'race,' and massively unequal divisions of income as factors impacting how gentrification plays out differently in Cape Town.

In this chapter we show that there are gentrification blueprints, models and policies, but that their transfer is not always North to South, West to East, global city to provincial city, urban to rural or even urban to urban, etc.; and they are mediated by context. The geographies and mobilities of gentrification are complex, contingent and sometimes contradictory, and can learn from analyses of migrating neoliberal governmentalities (e.g. Ong 2006; Ferguson 2009; Park, Hill and Saito 2012).

In this chapter, we make an important point that there is no clearly bounded or explicitly labelled 'gentrification *policy*' because governments, certainly in the global North, know that to label a policy as a *gentrification* policy' would be foolish given the negative associations of the term with social cleansing. What governments have done is to utilize policies with other more neutral labels (e.g. urban regeneration, urban renewal, urban renaissance) or with liberal moral discourses (e.g. mixed communities policy, urban sustainability) or the economically feel good term 'creative city'. Sometimes, new town projects under the label of 'eco-city' or 'smart city' are proposed and implemented, placing gentrification pressures on neighbouring districts too. Quite simply these 'gentrification policies without the name' soft peddle various ideas and programmes of gentrification. The very fact that no labelled gentrification policy exists makes resisting gentrification more complex and the process itself more 'backdoor' and indeed aggressive. Moreover, the unusual situation now is that ideologically quite different actors are picking and choosing bits from these models to announce their global urbaneness, but such ideas either fit uneasily with local politics and cultures or become remade to be acceptable to local politics and cultures. In looking for a gentrification blueprint, we must be clear that policies only occasionally move as fully formed things. More usually, what moves when policy is seen to replicate itself over time and across space is a far more disaggregated set of knowledges and techniques. Once these have moved then they get translated into policy, sometimes recognizable as the originating policy brand or type, sometimes not. This process is dependent upon highly contingent translations and innovations, and it is rarely linear. Sometimes, policies travel only as cosmetic make-up to enable the implementation of longer-term endogenous development projects that local elites have long envisioned.

In investigating a 'global gentrification blueprint', we contribute to extending understanding of what Brenner and Theodore (2002: 349) refer to as the relationship between city building and 'actually existing neoliberalism'. In doing so we elaborate on the diverse geographies of privatization driven by the neoliberal tendencies that are occurring around the globe via the inter urban, global movement of theories, policies and techniques (Larner and Laurie 2010: 218; see also Ward 2008; McCann 2008). We show the ways in which similar neo-liberal governmentalities produce (or do not produce) gentrification around the world (also see Salomón 2009, Brenner, Peck and Theodore 2010; Peck 2002); and how processes of gentrification are (or are not) embedded in the fast-paced, self-reflexive logics of inter-city policy adoption, circulation, and learning (McCann and Ward 2010; Peck and Theodore 2010a, 2010b; McFarlane 2011). In doing so, we acknowledge that neoliberalism is not an overarching colonizing force, and that local developmental politics interact with neoliberalism in a multifaceted and spatially and temporally uneven way. For instance, East Asian scholars have been debating the extent to which regional developmental states have been transformed into neoliberal ones, and it has been acknowledged that the characteristics of developmental states have persisted in one way or another in East Asian economies despite degrees of neoliberalization (see Park, Hill and Saito 2012).

Peck (2003) argues that neoliberalism has much of its origins on the periphery, not the centre, in Chile and Mexico in the early 1980s. Although designed in the North, neoliberalism was first tried out properly in Chile under a 1973 US-backed military coup and in Mexico in 1982 under a US–led rescue package; it was then pushed in other Latin American countries as a condition of loans or renegotiations of debt by the International Monetary Fund (IMF), the World Bank and the Inter-American Development Bank (IDB). As Betancur (2014) states – Latin America was exposed to neoliberal shock therapy while being tightly tied to the economies of the North. The result was that much of the region's capital shifted from manufacturing to finance and commerce. Capital focused on new projects for the middle/upper classes, commoditizing cities. But the shift was limited by underdevelopment in the region, informality and limited market capacity. Later it was seen in the privatizing politics of Thatcher and Reagan in the 1980s. Since then, neoliberalism has become the centrepiece of urban policy seemingly regardless of the politics of national governments. As Larner (2000) shows, neoliberal techniques of governance can be deployed in very different political contexts and projects. This necessitates a relatively open reading of neoliberalism, but not a reading that says that neoliberalism is everywhere.

In his discussion of the 'globalization of urban policy', Cochrane (2007) argues that contemporary urban policy is focused on 'accumulation through the built environment', that is property development, and the management of growing inequality in society through disciplinary tactics, like Smith's (1996) revanchism. Others have gone a step further and argued that urban policy is now exported globally as part of a class-based project that remakes cities around the world for the purposes of capitalist accumulation (e.g. Peck 2003). Our critical political economy approach follows on from this.

GENTRIFICATION/URBAN REGENERATION MODELS

The Spanish Barcelona Model and the Bilbao Model have both had a significant impact on urban regeneration policy and gentrification in cities world-wide. McCann and Ward (2011) argue that they possess a kind of representational power, operating as a transformative policy imaginary across different sites. Peck and Theodore (2010a) talk about the apparent paradox that the more policies flow/mobilize the more they are tagged to places like Barcelona.

The 'Barcelona Model' is probably the most famous example of global urban policy transfer and it is widely recognized as a prime example of 'best practice' in 'urban renaissance'. The 1992 Olympic Games provided a significant event from which to sell the supposed 'success story' of Barcelona's regeneration (Marshall, 2004), with its apparently successful combination of cultural strategies and urban regeneration to address social problems; a model that has now been critically rebutted (Degen and García 2012). González (2011) has shown how 'policy tourism', in the form of experts visiting Barcelona, contributed to the international diffusion of the 'model'. The Barcelona model, however, was/is not really a clear construct (Blanco 2009; Monclús 2003; Marshall 2000; Garcia-Ramon and Albet 2000), nor is it a singular model. Nevertheless, it promotes urban design (top-down but small-scale interventions to upgrade neighbourhoods), governance (strong leadership in promoting a city's image) and strategic planning (again top-down, e.g. roads etc.). The result of such strategies is that former industrial spaces and working-class neighbourhoods have been appropriated by the service and knowledge economy for both residential and productive use, producing gentrification.

The irony is that the Barcelona Model was ever considered as unique in the field of international urbanism when it is far from unique. Behind this exemplar of progressive values is institutional capital with relatively conservative concerns and negative consequences relating to social polarization and social exclusion, which have been constantly sidelined. Nevertheless, the chairman of Britain's Urban Task Force in the late 1990s, the 'starchitect' Sir Richard Rogers, adopted Barcelona as the model of British urban renaissance policy under the then New Labour government (see Lees 2003). Indeed the introduction to the Urban Task Force's *Towards an Urban Renaissance* had a foreword by Barcelona's socialist mayor Pasqual Maragall. Demonstrating a policy mobilities loop, Richard Rogers subsequently became a member of the Mayor of Barcelona's Urban Strategies Advisory Council. Through its urban projects, Barcelona is now a significant player in a global urban network marked by symbolic competition and mutual imitation. The Barcelona Model promotes the kind of gentrified, sanitized city that Zukin (2010) and others have described, with the same collection of contemporary architecture, urban museums, etc. Mimetically, each city that draws on the Barcelona Model seems to aspire to the same urban regeneration menu without paying close enough attention to the inclusions and exclusions in the model.

The entrepreneurial and technical knowledge of the Barcelona Model was exported to Latin America by the *Ibero-American Centre for Strategic Urban Development* (CIDEU in the Spanish acronym) as the strategic partner and technical consultant for local governments (Steinberg 2001). Rojas (2004) argued that many of the strategic plans for upscale urban redevelopment in central or historical areas in Latin America in the 1990s were inspired by Barcelona's success. Parnreiter (2011) has discussed the professional and city networks behind policy mobilities, including CIDEU, 'which was created to export the "Barcelona model" to Latin America (Salomón 2009) and has more than 100 member cities from Ibero-America, which all have applied the "Barcelona Strategic Urban Planning methodology"'. González (2011: 13f) notes, a 'unidirectional flow is particularly true for the case of Barcelona where consultancies have effectively 'sold' the model to mainly Latin American cities'. Indeed, this is one of the reasons that there have been assertions more recently about a specifically Latin American gentrification (see Janoschka, Sequera and Salinas 2014). But González (2011: 4) stresses that the Catalan capital is a node, rather than a starting point, in the 'space of policy flows' (also see Peck and Theodore 2010a), implying that Barcelona has been a receiver (mainly from the US and Canada) as well as a sender of strategic planning ideas and practices. This is important and underlines the fallacy of ideas about the linear, diffusionist, progression of gentrification. Similarly, Monclús (2003: 417) sees Barcelona's reputation as based mainly on the city authorities' capabilities 'to borrow, adapt and elaborate original syntheses relating to the most advanced formulae of international urban planning culture'. In fact, what is sold as the 'Barcelona Model' is, more often, 'a handful of models' (Marshall 2000: 315) whose nature depends on who the buyer is (González 2011).

A GLOBAL GENTRIFICATION BLUEPRINT? | 119

Figure 5.1: The Puerto Madero redevelopment in Buenos Aires, 2012 (Photograph courtesy of Ignacia Saona)

Barcelona planners were behind two examples of state-led gentrification in Latin America that emerged in the 1990s – both so called 'urban renewal' projects – the Malecón 2000 waterfront regeneration in Guayaquil in Ecuador and the Puerto Madero redevelopment in Buenos Aires (see Figure 5.1). Both were carried out under the prevailing neoliberal planning philosophies that widely failed to keep their initial promises of social mixing and the provision of public infrastructure (see Cuenya and Corral 2011). Experts from Barcelona provided the Buenos Aires management team with ideas and strategies for the Puerto Madero rehabilitation. But when the Barcelona experts delivered their strategic plan to the mayor of Buenos Aires in 1990, protests erupted over the lack of local participation in the project forcing the Corporation to establish the *National Contest of Ideas for Puerto Madero*. Community action groups and others have argued that the funds generated by the sale of public land for the Puerto Madero redevelopment could have been better invested in social welfare projects elsewhere in the city. Critics also complain that although it has produced a new landscape pertinent to world-city status, the area is separate from the rest of the city socially and economically and the masses have been excluded from the project.

By way of contrast, the Malecón 2000 waterfront redevelopment in Ecuador, one of the most extensive urban renewal projects in Latin America, is seen to be a success. In 1996, an urban team from Oxford Brookes University in England was invited to make a proposal to renovate Guayaquil's deteriorated riverfront as a large public space, following the idea of similar urban projects known in Guayaquil, such as Barceloneta in Barcelona and Bay Side in Miami. The Oxford Brookes team stated that the main goal was 'to create a large public space addressed to all the inhabitants of Guayaquil, without any distinction, which could re-establish the relationship of the city with the river, endure and serve as trigger to initiate a process of urban regeneration of the city centre' (Carbajal, et al. 2003: 20). The preliminary proposal and the proposed management of the project were based on studies of the management models of Barcelona, Bay Side and Puerto Madero – which Mayor Febres-Cordero supported. Successive surveys of the wider public have shown high satisfaction with the project; indeed, internationally, the World Health Organization (WHO) and the Pan-American Health Organization (OPS) declared it a 'Healthy Public Space'. It supposedly demonstrates urban regeneration without gentrification, that is direct displacement, as existing residents were allowed to remain in their houses; but there have been indirect displacements (see the case study in architectureindevelopment. org).

The new focus in Barcelona is the 22@Barcelona Project, which has shifted the focus away from tourism and onto the new technology industry. The 22@Barcelona model is the newest model to be exported from Barcelona to cities like Rio de Janeiro, Boston, Istanbul and Cape Town (see 22barcelona. com). The project began in 2000 on industrial land in the historical Poblenou neighbourhood of Barcelona.

This 'productive renewal' is committed to mixed use that fosters social cohesion and leads to more balanced and sustainable urban and economic development. The model being sold is the progressive urban and economic regeneration of industrial areas and a compact *and* diverse city. This model has also made its way to East Asia. Barcelona was highlighted as a model city for urban renewal in the 2010 Shanghai Expo whose theme was 'Better city, better life'. The Barcelona City booth for the Shanghai World Expo was designed to show how to define and reflect Urban Best Practices from the models of the Ciutat Vella and 22@ districts in Barcelona. The entire world was able to witness the only two urban regeneration projects mentioned. Indeed, the Barcelona Model has reached iconic status, as evidenced in a World Wildlife Fund India report where the executive summary asks: 'So, how can we ensure that India's future urban trajectory follows Barcelona rather than Atlanta?' (Sangal, Nagrath and Singla 2010: 7). In April 2013, an Indian government delegation comprising 30 city and public administrators from 20 different cities across India visited Barcelona. During the visit, the delegation had the opportunity to exchange views on policies and urban activities with representatives from the City of Barcelona. They visited Barcelona's 22@ district!

Like the Barcelona Model, the Bilbao Model is a regeneration model, although its focus is much more on arts-led regeneration (see Rodriguez, Martinez and Guenaga 2001). Vicario and Monje (2003) argue that '. . . although the urban regeneration strategies deployed in Bilbao are touted as unique, innovative and exemplary, in fact they are a rather recent continuation of a model first devised years ago by numerous cities in the US and the UK. Indeed, the intervention model followed in Bilbao was explicitly inspired by strategies developed earlier by cities such as Pittsburgh, Birmingham and Glasgow . . . Bilbao is, therefore, a significant example of the well-known approach dating from the 1980s where flagship property-led redevelopment projects are central ingredients of urban regeneration . . .' (p. 2384). The

Guggenheim Museum was opened in Bilbao, in northern Spain, in 1997; it was commissioned by an energetic mayor who hoped to turn the city's fortunes around. It is claimed that in the first three years after the museum opened it raised over €100m in taxes for the regional government, enough to recoup the construction costs and some more. By the late 1990s, debates concentrated on what became known as the 'Guggenheim Effect' (Gómez and González 2001) or the 'Bilbao Effect'. Many other cities, like Berlin, Abu Dhabi and Venice, have sought to replicate the supposed success of Bilbao, and the Guggenheim Foundation have been eager to meet that demand. Indeed, they have been accused of cashing in on their success and 'franchising their brand name', much like McDonald's. The new Guggenheim Abu Dhabi is to be located in the Cultural District of Saadiyat Island in Abu Dhabi, the capital of the United Arab Emirates (UAE), as a local offshoot of the Guggenheim and the Louvre. But there have been many issues around it, from rejection of American hegemony in Abu Dhabi to concerns over worker conditions. In a further twist, Serbia recently signed an agreement with an Abu Dhabi investor of a multi-million dollar Dubai-style riverside redevelopment – Belgrade Waterfront – marking one of the first forays of real estate tycoon Mohamed Alabbar and developer Eagle Hills and Emirates in Central/Eastern Europe. The plan is for 'a glass forest of hotels, office buildings and apartments for 14,000 people, the largest shopping mall in the Balkans and a curvaceous 200-metre tower on two million square metres of wasteland by the River Sava' (Sekularac 2015). Hundreds of protestors reportedly gathered as the agreement was signed concerned at the scale and cost of the project.

GENTRIFICATION POLICIES

We started this chapter in the global North/West (in Spain) on purpose, to demonstrate how easy it is to fall into the trap of assuming a diffusion towards the global South/East. We want to argue in this chapter that we need a much more nuanced understanding of how policies often dubbed as neoliberal were in fact existing already in many countries in the global South. Yes, there has been a rise in neoliberal urban policies over the past thirty years or so; indeed, almost all nation states have adopted neoliberal policies of some type or other (Harvey 2005), but this does not mean that states did not already have versions of these historically. In recent decades we have seen the election of right of centre governments in North America and Western Europe, stabilization policies in Latin America, the move to post-socialism in Eastern Europe and the former Soviet Union, and marketization in previously socialist countries like China. Neoliberal policies have gradually replaced the post-war ideas of Keynesianism with liberalization, deregulation, privatization, and depoliticization. In looking at policies that promote gentrification in the global North and in some parts of the global South, we can observe a shift in policy discourse from the manager role/welfare state role towards a more entrepreneurial role of the (municipal) government (Brenner 2009: 44). A policy discourse consists of: 'a specific ensemble of ideas, categorizations and narratives that is being produced, reproduced and/or transformed in policy practices' (Hajer 1995 in Arnouts and Arts 2009: 206). Policy discourse determines how policy is made and indeed affects the content of policy. It is made up of 'repeatable linguistic articulations, social-spatial material practices and power rationality configurations' (Richardson and Jensen, 2003: 16). Importantly, policy is not merely a derivative of neoliberal discourse (see Chapter 2), but co-produces this policy within the discourse (Arts, Lagendijk and van Houtum 2009). For example, the general principles of good governance and other moral considerations within national and local government ensure that the sharp edges of neoliberal discourse are trimmed. In other places, historically brutal and coercive urban policies are not dissimilar to neoliberal policies and as such befriend them.

Twenty-first-century gentrification has been, and is being, mobilized through three particular policies as we speak – zero-tolerance policing (a form of revanchism – see Smith 1996); mixed communities policy (or gentrification by stealth – see Bridge, Butler and Lees 2011) and creative city policy (or creative gentrification – see Peck 2010), the latter includes aspirations to the former – for creative types are said to desire social and cultural diversity/mixity. In this chapter we focus on these three *gentrification* policies without that name!

Zero-tolerance policing policies

Neil Smith (1996) discussed zero tolerance policies in NYC as part of his 'revanchist city thesis' (see Lees 2000; Lees, Slater and Wyly 2008: 222–34). But what is more interesting in terms of the themes of this book is his discussion of the revanchist city in relation to Third World cities:

> Third World cities have for a long time been scripted in the West as similar kinds of 'revanchist cities', cities where nature and humanity habitually take vicious revenge on a degenerate and profligate populace. . . . The organized murder of street kids in Rio de Janeiro, the Hindu massacres of Muslims in Bombay, the pre-election slaughter of South Africans in Durban . . . these and many other dramas present Third World cities to Western audiences not simply as places of extraordinary and often inexplicable violence but as places of inherently revengeful, perhaps lamentable, but often justifiable violence. (Smith 1996: 208).

Smith (1996) asserted that revanchism has always been there in the Third World, and it could be argued that the way neoliberal governments in the global South deal with social inequality and class conflict is by the imposition of police force and brutality. As such, instead of arguing, rather naively, that zero-tolerance policy moved North to South or West to East, it makes more sense to argue that in the 1990s New York City's zero-tolerance model became a blueprint for a renewed, more technological and brutal policy implementation in the global South, one that even attained legitimacy through its claims to learning from elsewhere.

William Bratton's zero-tolerance anti-crime model for New York City has been copied around the world and is readily associated with supporting gentrification, often 'cleaning up' areas as a precursor to gentrification. In New York City, Bratton and Giuliani created a bottom-up strategy dubbed 'broken windows'. In 1994 William Bratton was hired by NYC Mayor Rudolph Giuliani as Commissioner of the New York City Police Department (NYPD) and began to target 'quality of life offences': '[t]hese so called "beer and piss" patrols focused on drunkenness, public urination, begging, vandalism, and other anti-social behaviour' (Fyfe 2004: 45). Their 1994 strategy 'Reclaiming the Public Spaces of New York' was underpinned by George Kelling and James Wilson's (1982) 'broken windows thesis' (see Innes 1999: 398).

The success of the crackdown in NYC led to the adoption of zero-tolerance policing in the UK soon after, when Jack Straw, then shadow home secretary, visited NYC (Fyfe, 2004). In 2003, the ex-New York City Mayor Giuliani was paid a $4.3 million consulting fee for policy advice on creating a citywide crime programme in Mexico City, recommending stiffer penalties and similar crackdowns on minor offences to those he oversaw in New York City in the 1990s. He was invited to Mexico City by multibillionaire Carlos Slim (who paid most of the tab) and then-mayor Andrés Manuel López Obrador, who requested his specific expertise in urban cleansing so as to 'rescue' the 'crime infested' historic centre of Mexico City. Giuliani travelled through the Tepito barrio (internationally known for its dominating presence of informal vendors, known as *ambulantes*) in Mexico City with a caravan of 300 security agents and a helicopter soaring above. Giuliani Partners LLC, the former mayor's security consulting company (formed after he left the mayor's office), wrote a 146-point plan in 2003, known as 'Plan Giuliani' which attracted full government backing. The Plan set out to do the same in Mexico City as had been done in New York City, but the plan was criticized for ignoring cultural differences and on-the-ground realities such as widespread corruption and poorly trained police, an ineffective judicial system and the importance of the underground economy in Mexico City. Unlike in New York City, the Plan did not deliver a sense of stability and security (see Mountz and Curran, 2009). But one cannot talk about 'Plan Giuliani' without also discussing Manuel López Obrador's wider plans for Mexico City, and especially his Programa de Rescate – a global urban strategy that sought to gain Mexico City global city status. The idea was that this state-led policy would attract investment into the historic city centre (the Centro Histórico) and in so doing attract upper/middle-class residents to live there and tourists to visit (see Walker, 2008). This neoliberal municipal gentrification programme, spearheaded in 2001, tallied with 'Plan Giuliani' in wanting to take back (gentrify) the Centro Histórico of Mexico City from ambulantes. A large-scale project, the Programa de Rescate, aimed to renovate an area three times the size of the historic area of Barcelona. The process of gentrification in this example was/is a three stage process: first, replacing the water and sewage infrastructure and building a commercial corridor; second, building hotels, a visitor centre and skyscraper; third, and most visceral, removing and relocating the 30,000 or so ambulantes who live and work in the Centro, adding public amenities and increasing security (including panic buttons at different sites linked to the police). In Mexico City we see here a form of state-led gentrification as a form of roll out neoliberalization in which the state has implemented zero-tolerance urban policies to gentrify 'problem' areas that were created during an earlier period.

Similarly in Brazil, Rio de Janeiro's police have enacted zero-tolerance policies, declaring a new war against the city's minor offenders and petty criminals (the war of course is historic and police violence is not new in Brazil, as the 1980s slaughter of street kids attested to). Again, they credit ex NYC Mayor Giuliani for the idea. After reading a news story about the 'success' of Giuliani's zero-tolerance policing in cleaning up NYC, Military Police Chief Colonel Celso Nogueira concluded that the strategy was well-suited to Rio's densely populated, heavily touristed, neighbourhood of Copacabana, and he began a pilot programme there. They increased the number of beachside surveillance cameras and radio-patrol police; cracked down on unlicensed vendors, petty thieves, and unruly motorists; and increased the arrest of those who were coercing drivers to pay for free parking. The programme was then expanded to discourage beggars' encampments and place street children in shelters, unfairly targeting the city's poor and vulnerable. Unable to forcibly evict street people, Rio's police put pressure on them by removing their bedding materials from the streets and increasing sweeps on street vendors and other 'marginals'. In the run up to the 2014 World Cup, police violence increased, including beatings and the use of pepper spray. Advocacy groups and public officials have said that the programme victimizes Rio's most defenseless, cleaning up streets for tourists and the well off. But unlike Mexico's 'Plan Giuliani', Rio's Zero Tolerance is far smaller in scale, focused on specific neighbourhoods. It lacks fully-fledged government backing and has attracted some community-based council support. In late 2009, Giuliani, however, announced that they had a security consulting contract with Rio de Janeiro regarding the 2016 Summer Olympics! (also see Chapter 7 on the Olympic-related escalation of large-scale redevelopment with hostile state actions in Seoul; see also Shin and Li (2013) for the exclusion of migrants during the Olympic Games period in Beijing).

Mixed communities policy

Encouraging socially mixed neighbourhoods and communities by bringing middle-income people into low income neighbourhoods has become, and for the moment is continuing, as a major planning and policy goal in North America and in a number of West European countries. The largest and most heavily funded programmes have been in the US, the UK and the Netherlands, but it has also been policy in Canada, Ireland, France, the Netherlands, Belgium (see Bridge, Butler and Lees 2011) and in Anglo-American territories e.g. Australia and Puerto Rico. Given that too strong a dose of neoliberalism is political suicide, neoliberal urbanism began to be soft peddled in the 1990s and 2000s through the use of 'soft' discourses like social mixing, social capital, social cohesion, diversity, sustainability, participation and empowerment. These soft discourses are seen as neoliberal, but moral, commonsense. These soft discourses have been rolled out in place of redistributive policies, in a retrenching welfare state, using strong neo-liberalism (see Jessop 2002a; Mayer 2003; Sheppard and Leitner 2010).

But they have not been rolled out everywhere, for in some parts of the global South mixed communities policy is not new at all. In Kolkata, India, social and economic mixing has been in place since the state housing board was set up in 1972, well before 'mixed communities' became policy in Western cities. In 1972, the state government acquired 1,025 acres of land for the Bihar State Housing Board so that an urban housing project could be developed for lower-middle-and higher-income groups. This was a popular strategy that enabled an impoverished government to cross-subsidize between income groups. More recently, the state has a mixed-income housing strategy – a policy enacted to regulate all state built and new public-private partnership schemes that is tied to the government's new rhetoric on affordable housing (see Sengupta 2013). But like in the West, the approach often inhibits the sustainable mix of communities. More recently market liberalization has aided the state-aided construction of luxury housing estates that are densifying and gentrifying the city's fringe and also creating a gated, bourgeoisie territory of power and aesthetics that is marketed to affluent buyers as a good place that guarantees global services and lifestyles. Socio-culturally these new estates show rising incomes and decreasing family sizes and have resulted in the indirect displacement of lower-income residents due to increased rents and costs. Loopholes in means testing and the allocation of low-income homes shows signs of upper-income invasion, showing that mixed-income policy is producing gentrification. As such, lower-income households are now rapidly declining in the state provided/assisted housing in the suburbs. What is happening on Kolkata's fringes is symptomatic of wider transformations in other cities in India and indeed the global South. The centrality of neoliberal development is quite apparent in the process and is leading to an uneven and inconsistent gentrification in the suburbs, aligning with gentrification trends in the West. Bose (2014), writing on India, is quite clear that a form of global gentrification is taking place, through which people and place are being fundamentally restructured.

The concept of 'mixed communities' or 'social mix' was not new in the global North either (Lees 2008). Indeed it re-emerged in the 1990s in reaction to the large concentrations of socially homogeneous populations of poor people residing in the inner cities of Western Europe and North America. The US Department of Housing and Urban Development (HUD)'s HOPE VI programme (Home ownership and Opportunity for People Everywhere) of poverty deconcentration was passed by Congress in 1992. This programme has begun to demolish large public housing projects at the centre of US cities and to disperse (*displace*) the project's residents. In their place, they are building new mixed income communities. The notion of mixed-income communities offered similar arguments to those made in the 1950s around 'balanced communities', the idea being that social diversity will enrich the lives of residents, promote tolerance of social and cultural differences, and offer educational and work role models. HOPE VI, however, is under increasing criticism for not delivering a truly mixed community nor the benefits supposed to come with this new mixed community (see Joseph 2006; Fraser and Nelson 2008). As Wilson (1987) made clear, reintegration with the middle class alone is not enough to generate social mobility. There needs to be a change in the structural economic conditions that inhibit access to employment, etc.

Evidence from the global South comes to similar conclusions. Drawing on a comparative analysis of Chicago and Santiago de Chile, two historically segregated cities, Ruiz-Tagle (2014) has assessed the relationship between neighbourhood social diversity and several dimensions of socio-spatial integration. Arguing against the consensus of mainstream policymakers in both countries, who strongly believe and defend the idea that the mere physical proximity between different social groups leads to social integration, Ruiz-Tagle finds no evidence that social-mixing housing policies defeat the barriers of class-oriented fear and exclusion. The author discovers that the physical proximity of different social groups, irrespective of the urban processes that bring them together, does not directly create the outcomes that supporters of poverty deconcentration policies believe. In other words, neighbourhood social diversity is not a sufficient condition for enhanced opportunities, better intergroup relationships or less exclusion from the housing market. In fact, Ruiz-Tagle argues it might well be the opposite, as lower status groups that live in mixed neighbourhoods and that were researched in Santiago and Chicago showed limited job opportunities, limited access to quality education, highly difficult intergroup relationships with upper status groups, and still suffered from exclusionary housing.

The HOPE VI Programme of state-led gentrification has been rolled out in Puerto Rico (Fernández Arrigiota 2010). Given its political and financial links to the US federal government, this is perhaps not surprising. Indeed, the public housing system is mainly financed with programmes from HUD. Like in the US, disinvested public housing is being demolished to make way for market rent housing, and tenant-based vouchers are being used for poverty de-concentration. One example is the high-rise public housing estate of Las Gladiolas located in the financial district of San Juan, Puerto Rico. Home to 670 families, it was built in 1973 as part of Puerto Rico's slum (barrios) clearance initiatives at that time. Like high-rise estates in other parts of the globe (see Lees 2014b on the Aylesbury Estate in London), Las Gladiolas began to be stigmatized as a failure – crime and drug ridden, delinquency etc. (see Figure 5.2). The Housing Department announced plans more than a decade ago to

Figure 5.2: The Aylesbury Estate in London, 2012 – the *Daily Mail* stigmatized it as 'Hell's waiting room' (Photograph by Loretta Lees)

demolish the estate. Significantly, Las Gladiolas was located just a block away from the Golden Mile financial district, in other words on high value land. Mano Dura, the zero-tolerance approach to crime that Police Superintendent Pedro Toledo implemented under the Rosselló administration from 1992 to 2000, targeted public-housing projects as the areas where most criminal activity in Puerto Rico took place, and the redevelopment of Las Gladiolas was sold to the public as crime reduction (Fernández Arrigiota 2010). The Puerto Rico Public Housing Authority obtained a demolition permit from HUD in 2006, and that same year 200 Las Gladiolas families filed a lawsuit in federal court challenging the eviction order, arguing they were notified without due process and requesting they be allowed to participate in the area's development. The plaintiffs charged that the commonwealth government failed to consult residents on plans to raze the complex – which dated back to 2000 – and had not kept the buildings in livable condition in accordance with federal law. But Las Gladiolas was demolished in 2011, and the site was redeveloped as townhouse-style subsidized housing. Here, state-led gentrification can be seen to be a postcolonial 'project' that has discriminated against Las Gladiolas' residents. The policy itself could be seen as an example of US imperialism in Puerto Rico (see Briggs 2002).

Mixed-income housing development with the objective of redevelopment, integration along racial and social lines, and the alleviation of urban poverty and segregation, is also being undertaken in South Africa. In Johannesburg, mixed-income housing is being promoted to denounce the negative perception that the poor and rich cannot live side by side and that public-private partnerships in housing development work (see Onatu 2010). To deal with informal settlement formalization, the local government has embarked on a massive mixed housing development – Lufhereng in Soweto – a large-scale, mixed-income and mixed-tenure housing development which will comprise 24,000 mixed-income houses, with schools, clinics, sports

Figure 5.3: Las Gladiolas, Puerto Rico, 2007 (Photograph courtesy of Melissa Fernandez)

fields and recreational facilities making up a complete, sustainable community. The name Lufhereng is derived from a Venda word 'lufhera' and a Sesotho word 'reng', which together refer to a place where people come together with a united commitment. Shaw (2011) argues that while gentrification may be an 'elastic yet targeted' process of what is 'now global' . . . social mixing policies are not, but the cases here indicate otherwise.

Creative city policy

Richard Florida's (2002) creative city thesis recasts urban competitiveness between cities as cultural and economic 'creativity'. This thesis has travelled around the world; through talks that cities have paid to hear, conferences, promotional activities, and policy blueprint documents.

The mobilization of the 'creative city' has become big business with creative city consultancy growing. Creative city consultants sell blueprints of 'how to create' a new and exciting urban environment (live-work units, bike paths, historic architecture, etc.) that will attract creative types, is very similar to the type of urban environment that pioneer gentrifiers in Western cities sought in the 1960s and 1970s – sustainable, diverse, emancipatory, vibrant, on the edge, and so on. Creative city blueprints are quite simply a gentrifier's charter (Lees, Slater and Wyly 2008). More recently, however, there has been discussion of the global divergence of creative city policies (Cunningham 2009) and some scholars are pushing for a wider understanding of creative city discourse beyond its Euro-American cradle (see Pratt 2009: 19; also Luckman, Gibson and Lea 2009).

Why the creative city became so mobile is highly disputed. Some argue it is a result of the globalizing economy, part of the on-going restructuring of the economy (Prince 2010). Others like Peck (2001, 2005) argue that Richard Florida's creative city/class fits the neoliberal development agenda that is based on competition, gentrification, middle-class consumption and place marketing perfectly: 'The banal nature of urban creativity strategy in practice' is covered up with Florida's sales pitch 'in which the arrival of the Creative Age takes the form of an unstoppable social revolution' (Peck 2005: 740–1). According to Peck (2005), the field of urban policy has lacked new, innovative ideas for a longtime and therefore creativity strategies became popular very quickly because they delivered 'both a discursively distinctive and an ostensibly deliverable development agenda' (p. 740). It seems as if policymakers just accept the assumption that creativity is the basis for economic growth without suspicion (Peck, 2005). Within the longue durée of capitalist expansion, the diffusion of creative policy matches the shift to the consumption of images and spectacles that marks the exhaustion of the present accumulation regime as a productive force (Ross 2007 in Prince 2010: 121).

Cunningham (2009: 376) has looked at the different ways in which it is used around the world. He says that the discourse should be 'thought of as a Rorschach plot, being invested in for varying reasons and with varying emphases and outcomes'. Critical geographers have stated that the creative industries are in fact the latest step in the commodification of culture, as already described by Adorno and Horkheimer in 1948 (O'Connor 2009; Prince 2010). The creative city and its related concepts are used by a neoliberal state apparatus, that is: 'intent on creating markets that can govern culture through their particular incentives and constraints, while opening new territories for capitalist exploitation' (Prince 2010: 120). This is certainly the case in East Asia where the creative city has been 'ideologically anointed or sanctioned' (Peck and Theodore, 2010a: 171) by states like Singapore, and China. These are fascinating case studies given that in China and Singapore creative city policy sits uncomfortably within authoritarian states in which neoliberal institutions coexist. Indeed, Singapore is now 'selling' its creative city model to South America, Dubai and Pakistan (see Kong 2012, on creative city policy in Singapore).

Kong and O'Connor (2009) argue that much of the thinking on cultural and creative industries in national and city policy agendas in East Asia has been derived from the European and North American policy landscape. China now has six cities listed in UNESCO's Creative Cities Network – Shenzhen, Chengdu, Shanghai, Beijing, Hangzhou and Harbin. No other country has as many internationally acclaimed creative cities. The creative industries in China stand in the face of 'old industrial' China – its factories and 'Made in China' label. But as O'Connor (2009: 175) wonders:

> The issue of creative industries is part of the question of China's future – its relationship of difference from and similarity to the

West – as it is also a question of China's past – what influence will that past have on its future trajectory, is it a resource for, or burden on, this future?

The first city in China to join the then mostly Western league of creative cities was the southern city of Shenzhen in 2008. Both timing and geography were important in the development of Shenzhen as a creative city. Creative designers moved to Shenzhen in the early 1990s when the city was undergoing sweeping changes due to economic reforms and China's opening up policy. The Open and Reform Policy planted in the Shenzhen Special Economic Zone 30 years ago, which led to rapid development, enabled the growth of creativity in Shenzhen. Creatives gave industrial products a brand new image and pushed them up the value chain. Shenzhen is now a leader in graphic design, fashion and architectural designs, interior, packaging and industrial designs. This shift in city governance from 'Made in China' to 'Created in China' in Shenzhen has actively shaped its urban form and developmental model. Its geographical location, away from political centres of control and near to Hong Kong with its links West, was/is important.

In 2004, OCT Real Estates Co Ltd announced its plan to renovate factory buildings abandoned in the 1990s, into a new modern art and cultural centre at a cost of more than 30 million yuan (US$3.9 million) (News Guangdong 2007; see also O'Connor and Liu 2014). According to its CEO, the plan was to transform it into a district much like New York's Soho, while preserving its original appearance (News Guangdong 2007). They dispatched experts to visit Yaletown, the loft area of Vancouver, Canada, which allows people engaged in artistic activities to have an exclusive use of the buildings (ibid.). They copied Vancouver's Yaletown in OCT-LOFT. In 2012, URB (the Urbanus Research Bureau) organized a workshop in Shenzhen titled 'The Making of a Creative City – International Workshop on post-industrial development of Shenzhen' (Moving Cities 2012). One of its research projects was 'Creative City Shenzhen'. Their aim was to analyse the creative park model, advocated throughout the world. Part of this was to research the creative industries that had settled in the OCT-LOFT and also the social capital of the creative class – their social networking relationships, needs, perceptions and beliefs. The city aims to make the cultural and creative industries its fourth industry along with the high-tech, finance and logistic industries! As Prince (2010: 121) argues the rapid diffusion of creative industries policy is the result of so many policymakers, activists, council and government officers, cultural entrepreneurs, researchers, and academics from so many places being able to incorporate the concept into their political, cultural, economic and social projects.

In prioritizing attracting the creative class to Shenzhen, there has been a social restructuring in which white collar technocrats have increased and blue collar workers declined. This is not class replacement but class displacement in that Shenzhen's factory jobs have been moved elsewhere, and the City now actively excludes manual labourers, sanitation workers, and other low-end migrants from transferring their hukou to Shenzhen. Disturbingly, this process of, dare we say social eugenics, is being reinforced by the demolition of centrally located urbanized former rural villages (a vestige of earlier planning regimes) housing low-income migrants in particular, in a process of gentrification. As Booyens (2012) argues, creative city policies are gentrification policies because creative urban renewal exacerbates existing inequalities and marginalizes poor people. They uproot lower socio-economic residents and businesses and create social polarization. Cunningham (2009) has argued that in the global South policymakers tend to link creative industries with poverty alleviation, basic infrastructure development, social inclusion and the promotion of cultural heritage and diversity, but this has not been the case in Shenzhen.

CONCLUSIONS

These state-led urban projects around the globe verify Smith's (2002: 446) assertion that urban regeneration processes 'represent the next wave of gentrification, planned and financed on an unprecedented scale'. Betancur (2014: 2) has found that gentrification in Latin America has not emerged from the local dynamics of restructuring; rather 'it has been launched by governments with the assistance of international agencies pushing restructuring to create new markets and advance formulas of urban competitiveness in the South'. He argues that colonization and neocolonization have shaped Latin America into a single polity subject to the mandates of the Washington Consensus. In the twenty-first century, a global gentrification has been launched by governments. However, the emergence of gentrification in different places is embedded in very different stories, the Washington Consensus has less influence on China, for example; but the impact of gentrification is the same everywhere. Peck and Theodore (2010a: 173) argue that '[e]ven the "same" policies tend to be associated with different effects in different places, by virtue of their embeddedness in, and interactions with, local economic, social, and institutional environments'; yet, this is not the case for gentrification policy which has the same core effect everywhere – displacement of poor/low-income groups in favour of the wealthy.

As Inzulza-Contardo (2012) has said, recognizing urban regeneration as gentrification is important, for in Latin America, like elsewhere, terms such as 'regeneracion urbana' (urban regeneration), 'renovacion urbana' (urban renewal) or 'mejoramiento de barrio' (neighbourhood improvement) are used in official plans and policies – there is no mention of gentrification! This is a recurring situation in Asia too, where gentrification as a process exists ontologically, while seeing its epistemological absence in public discourses (see Ley and Teo 2014; Shin and Kim 2015). These plans and policies must be recognized as 'gentrification', and protective legislation is needed in countries around the world to mitigate what Roy (2005) calls the 'unintended consequences' of these blueprints – in the case of gentrification, displacement of the poor. How can gentrification be a good model or policy when it is to the detriment of the less well off in society?

McCann (2004) argues that this serial reproduction of policies, as Harvey (1989b: 10) referred to it in his entrepreneurial city thesis, or 'policy transfer' (see Dolowitz and Marsh, 2000), tends to foster weak competition and crowding in the market place 'that works to the detriment of most cities by fostering a "treadmill" effect in which every city feels an external pressure to upgrade continually its policies, facilities, amenities, and so on, to stave off competition and maintain its position in the competitive urban hierarchy' (p. 1910). It is hoped that cities and governments around the world will soon wake up and realize that models and policies that produce gentrification (Smith's 2002 gentrification generalized) are not a solution but rather a long-term problem. Urban policy needs to be made responsible at a global scale (see Massey 2004, 2007, 2011). It is high time that alternatives to gentrification were developed by policymakers and planners; this will necessitate not just some real creative thought, but also the courage to do things differently, to be leaders not followers.

6 Slum Gentrification

In the twenty-first century, slums are seldom depicted as urban 'narratives' of capitalist exploitation (as they were in 1872, when Engels published *The housing question* as an account of speculative landlordism by the industrial capitalist class): instead, they are currently often referred to, by NGOs or transnational institutions such as the United Nations, as an effect of the state's incapacity to provide shelter or the state's deficient land planning capacity. Place-specific conflicts further escalate the rise of slums, as has been observed in South Africa's post-apartheid spatial politics disadvantaging the poor (Murray 2009) or in Zimbabwe where the state's deliberate suspension of informal housing resulted in concentrated demolition in towns that were strongholds of the opposition party (Potts 2011). In fact, the United Nations Habitat report indicates that the highest shares of slum populations in national populations are found in countries emerging from conflicts, often found in Africa (UN Habitat 2013: 21–2). With regard to cities of the global North today, theoretical debates on slums are almost non-existent despite the global North still having slums (see Ascensao, 2015). This book gives us the space to undertake much needed comparative discussions on this matter. This is especially important, as the global effects of the 2008 financial crisis in many countries weaken or erase welfare states and reduce public shelter provision and capacity, even in formerly rich areas of Western Europe and North America.

Currently, an increasing array of literature – whether part of gentrification discussions or not – has brought the redevelopment of slums back into global debate (e.g. *Beyond the Return of the 'Slum' in* CITY 2011). We believe that efforts to introduce gentrification narratives into slums around the world are useful for both decentring Northern gentrification theory and contributing to local discussions on slum redevelopment. Here, we focus our attention on how such spaces depicted historically (incorrectly or correctly) as slums are being increasingly subject to capital (re)investment. This condition means that Southern slums can no longer be conceptually excluded from gentrification debates.

Mike Davis (2006a) argued that slums were one of the leading forces of urbanization in the world, and described them as underserviced, heterogeneous and concentrated spatial manifestations on environmentally risky land, containing 'surplus humanity'. The conventional wisdom is that the world now has about one billion slum dwellers. The usual suspects for hegemonic slum narratives still lie in the urbanizing countries of Africa, Latin America, South Asia and Arab countries. Currently, in Durban, 23 per cent of the urban population live in shacks (800,000 inhabitants) and one million live under the constant threat of eviction; in Karachi, Pakistan, half of the population live in informal settlements (5 million); in Istanbul, Turkey, 80 per cent of the city's housing stock is estimated to be 'informal' (1.8 million); in Greater Buenos Aires there are 120,000 people living in informal settlements (Cabannes, Yafai and Johnson 2010); in Morocco, 33 per cent of the urban population live in informal areas, which until recently have been tolerated by the authorities, until the early 1990s when policies of massive slum eradication started (Bogaert 2013). The slum problem is still critical to societies and governments world-wide. And the devalorized category 'slum' offers potential for large-scale capital speculation in the form of gentrification.

According to the UNDP 2014 Human Development Report, income-based measures of poverty show that 1. 2 billion people in the world live on USD$1.25 or less a day and almost 1. 5 billion people in 91 developing countries are living in poverty. The actual situation though is starker because almost 800 million people are at risk of falling back into poverty as a result of war, crisis, etc. This pessimistic future brings the unsolved housing question back to the fore. In a world where more than 50 per cent live in cities, macroeconomic change, poverty, urban centrality and redevelopment opportunities are intertwined variables that need to be addressed together. And given the fact that even wealthy cities in the US and Europe have not succeeded in erasing their own slum neighbourhoods (Samara et al. 2013), comparative work on slums globally is much needed. In 2006, the megalopolis of Los Angeles held the biggest slum in the developed world, with 100,000 people concentrated in tents (Davis 2006b). Ascensao (2015) also discusses several cases of informal and illegal *clandestinos* which settled around Central Lisbon in Portugal during the 1980s, in enclaves that were (then) tacitly accepted by the state. Recently, these 'slums' have become targeted for 'gentrification', through modernizing state rhetorics, populations are being evicted and slums demolished.

There is a whole politico-economic dimension to the class-related effects of the redevelopment of informal settlements in the world, as Desai and Loftus (2013: 790) discuss regarding Mumbai:

> speculation [in informal or slummy areas] is accentuated by ostensibly philanthropic acts when seeking to improve tenure security through investments in infrastructure of slum areas by NGOs, developmental agencies and governments. Such investments are in danger of becoming conduits through which an emerging (albeit still nebulous) landlord class are able to exploit the need to switch capital from a primary to a secondary circuit.

This is another example of what we discussed in Chapter 3 about the speculative switching of capital at the planetary scale, and Desai and

Loftus (2013) note the increased ability of landowners and land/property renters to profit from slum insecurity. Like with the use of mixed communities policy to gentrify 'slum' public housing in the West (see Bridge, Butler and Lees 2011; see also Chapter 5), infrastructure projects or upgrading policies in slum neighbourhoods in the global South are also hinged on the 'false assumption' of trickle-down and the redistribution of positive externalities among local residents. And like in the global North, these policies are enacting 'gentrification by stealth', from Mumbai to Rio de Janeiro. Often this is related to the discursive strategies of world-city making and their associated policies, also discussed in Chapter 5, which more often result in displacement and dispossession (see Goldman 2011: 59 on Bangalore's transformation into 'a world-city nightmare'). The main questions we ask ourselves here are: does the 'gentrification lens' offer something new in terms of understanding the current nature of slum change and redevelopment happening around the world? Can the new processes of change or power exerted in slum areas be deemed gentrification, or is it rather diversified forms of slum upgrading? And who are the key actors in this type of gentrification? How widespread is it? What is the significance of strong property speculation in places like Mumbai or Beirut when juxtaposed with traditional discussions about slum settlements, eviction and infrastructure development? And maybe more importantly, can we learn something new about resisting gentrification from slums in the world?

THE GENTRIFICATION DEBATE IN LIGHT OF CONTEMPORARY SLUM CONFLICTS

The word 'slum' is polysemic and in some places non-existant. This can be a hurdle when using it as a single analytical category. In the non-English speaking South, the term is barely known. Often 'slum' or its translated equivalent, perceived as a negative name by people, is rarely used. Many people prefer instead terms such as *barrio* (neighbourhood), villages, settlements or community (Cabannes et al. 2010: 15). Also, by using 'unauthorized colonies' instead of slum (Lemanski and Lama-Rewal 2013), one can see the far richer and more complex processes of social stratification existing in many of these urban spaces around the world. The use of slum as a concept might well be misleading and epistemologically problematic as it can offer nothing more than a blurred depiction of stigmatized urban areas. Indeed, the term 'slum' is often used to justify the erasure of important low-income areas targeted for market-led redevelopment. But other authors like Arabindoo (2011) claim that use of the term is not essentially negative if it leads to better geographical understandings of the many aggressive ways socially produced urban spaces are becoming absorbed by neoliberal domains. We use the term 'slum' in this chapter, but in quotes, indicating our unease with the term. We recommend whenever possible other labels such as favela, barriadas, poblaciones callampas, bidonvilles, clandestinos, villas miseria, gecekondu, katchi abadis, panjachon, umjondolo, pirate subdivisions, unauthorized colonies, etc. (see UN Habitat 2003: 9–10 for examples of various expressions of slums around the world).

It is evident that a multidimensional definition of 'slum' is needed, but a working definition can be based on the most basic characterizations of overcrowding, poor or informal housing, inadequate access to safe water and sanitation, and insecurity of tenure (Davis 2006b). What applies to 'slum' in one world area may not do the same for another area. Dwellers of most *poblaciones* in Chile have gained formal land tenure and access to water and sanitation in the 1970s and 1980s, but they still experience severe problems of overcrowding and informal (usually dangerous) poor construction of their built environment (López-Morales and Ocaranza 2012), and so these enclaves remain as unplanned inner city territories, entrenched by the rapidly expanding forces of private redevelopment. Similar trends can be seen in the *villas* of Buenos Aires (Herzer 2008) or *barriadas* in Lima, but a totally different story can be told in Istanbul's *gecekondus* or Cairo's peripheral shanty towns (Elshahed 2015) where peripheral desert areas that the poor have traditionally used are currently experiencing new-build developments.

Histories, sizes, scales and locations, vary from case to case. There have been very large slums in Karachi for more than 120 years; and old (over 70 years), large-scale slums can be found in Buenos Aires, Argentina too. Large slums have also been created recently in Santo Domingo and Kabul. Some slums occupy vast peripheral, underserviced areas while others are located in the very centre of cities, or at least occupy an important part of the expansion areas of CBDs (e.g. Mumbai) or upper-income residential neighbourhoods (e.g. Sao Paulo, see Burdett and Sudjic 2011). Slums may come to occupy central locations if their initial peripheral nature is overshadowed by urban expansion. Some slums face the threat of state-led evictions or are constantly harassed by the police or mafia squads. Yet in other places, slum dwellers may have land titles and the forms of displacement are more subtle and less obvious like in Santiago de Chile, analysed below, where land economics and real estate markets play important roles in the radical changes occurring as developers acquire land for their projects; or like in Mumbai, where some settlers themselves become property speculators, and in many ways, play the role of gentrifiers.

As Samara et al. (2013) claim, neoliberalism has increased the socio-spatial divisions between groups of residents who become increasingly foreign to each other, and to other places in the city:

> What is different [now] is the size of the marginalized multitudes and the extent of their deprivation, neither of which can ever be entirely escaped no matter how many layers of security and distance more affluent groups put between themselves and everyone else. For formal policy and governance to the rhythms and disjuncture of daily life, the reality

of the divided city infuses local politics and culture. The ways in which it does so, and the consequences, will vary greatly between cities undergoing massive, planned redevelopment, such as Shanghai at one end of the scale, to those like Luanda and Managua on the other, where the local elite are too small, and the scope of transformation too narrow to sustain more than connected enclaves in what is otherwise (perceived to be) a hostile terrain. (Samara et al. 2013: 7)

Urban division remains, as it has for centuries, yet there is nothing 'natural' about this division. But maybe something new is going on today, compared to several decades ago: for the threats of demolition now encompass much larger areas, there are more rampant forms of economic coercion on settlers, stronger political violence, and so on (see Chapter 7). In addition, displacement also plays a central role in the commodification and financialization of hitherto unplanned cities (Desai and Loftus, 2013); this reflects not only the result of an imposed neoliberal logic but also the interplay between local and extra-local forces that constitute the 'substance' of neoliberal globalization. Finally, the more affluent interest groups seek to create and insolate their globalized enclaves as much as possible from the perceived chaos around them (Álvarez-Rivadulla 2007).

In a neoliberalizing world, GINI indices are usually higher in countries experiencing rapid and recent market-led expansion. The concept of the 'divided city' appears as the preeminent urban form for these and many other cases. In addition, an intangible vision of an 'inclusive city', very popular among policymakers, planners, officials and scholars, often produces quite the opposite effect, the deepening of social and spatial divisions (see Bridge, Butler and Lees 2011). However, the formal-informal divide is essentially a perceived one, albeit a powerful alibi for neglecting, transforming, and manipulating both parts of the city which are connected economically, socially, culturally and politically (Bogaert 2013).

The 'making' of divided cities is accomplished by a variety of means, but in the global South, it too often encompasses displacement, demolition and state violence, related to the introduction or intensification of commodified and financialized land and housing markets. Agents of the state, of private security or more shadowy actors play a key role here, often operating as proxies for particular coalitions of interests linked to accumulation and class (Samara et al. 2013). There are cases where slum gentrification simply means slum removal, but in other cases, it means the gentrification of slums in situ by wealthier in-movers, as has happened in the case of Mumbai that we analyse next. All in all, slum gentrification is a significant part of the urbanization processes going on in the global South. It is escalating, and is evident in cities like London in the global North too (Lees 2012).

SLUMBAI OR THE GENTRIFICATION OF ASIA'S LARGEST SLUM

Mumbai, Delhi and Bangalore are currently the most dynamic cities in India with rapid economic growth, and facing for the first-time privatization of their urban land and housing. Mumbai, the IT and financial capital of India, has an estimated 19 million people, but 65 per cent of its inhabitants are employed in the informal sector. Indian slums are of two kinds: the authorized ones, where the municipal government provides basic services, and the unauthorized ones, that represent 60 per cent of cases, and for which the municipality has no obligation to provide services and that are constantly under pressure of demolition. In these latter areas, densities can reach 820 people per hectare (Burdett and Sudjic 2011). The city of Mumbai is changing so rapidly and in such unprecedented ways that according to Harris (2012), it has become a 'prototypical' twenty-first-century city, even if there is global political concern about its process of under- and re-development. Mumbai has become fashionable across popular and academic Anglophone contexts over the past decade, but this discourse of the global city of Mumbai does not fit well with its 2000 slums and its 200 informal areas targeted for social eviction (Arabindoo 2011). There remains a mismatch between representational Mumbai and real Mumbai (Harris 2012).

Dharavi (see Figure 6.1), infamously known as 'Asia's largest slum', covers 212 hectares with a condensed mixture of small businesses, manufacturing workshops and sweatshops, schools, churches, amid hundreds of thousands of mid-rise and low-rise dwellings; different castes, languages, and ethnicities all mix in an environment where local and international planners have very often withdrawn from attempting inclusive, upgrading redevelopment plans (D'Monte 2011). Slums in India are not unregulated urban systems but deregulated ones, where

Figure 6.1: Dharavi Slum, Mumbai, 2010 (Photograph courtesy of Paroj Banerjee)

parallel rules of construction and service provision coexist and imply negotiations with mainstream legal frameworks and institutional powers. Residents invest in improving their dwellings and neighbourhoods, while they use political arrangements and negotiations with local officials to obtain informal and formal access to services as well as temporary *de facto* tenure, even if the latter are never completely guaranteed. On the one hand, politicians and other state authorities encourage squatting in exchange for electoral support, and on the other hand, usurp land for developmental purposes (Doshi 2015).

Although Mumbai's unauthorized slums all too often feature faulty electricity lines with no water or sewage supply, shacks were sold in 2011 for between USD38,000 to USD50,000 (Lees 2014a), confirming them as vibrantly emerging land markets. But who are the actors that are imposing redevelopment on currently occupied slum areas, and how are they materializing their power? Is Mumbai's redevelopment-led gentrification really showing what Harris (2008) calls the key site in the creation and promotion of the neoliberalization of South Asia?

Doshi (2015) recounts how new state formations consolidated middle-class power, and elite and middle-class participation in urban operations infused governing regimes with new mechanisms and 'logics of enclosure'. Currently displacement-inducing projects do not generate direct gains for specific groups but support an elite-biased world-class urban aesthetic. And this is happening at the same time as the power of locally elected municipalities, that the working classes have thus far used to access their needs, is in decline. A type of gentrified (some claim, right-wing) governance that varies across and within Indian cities has emerged. This ranges from new legal mechanisms and fast-track access to state bodies for developers and corporate and financial elites, and middle-class residential groups, to highly centralized bureaucratic urban development agencies, with middle-class residents' associations filing the vast majority of litigations.

This form of governance is not just about politics but strongly about class conflict. Middle-class activists demand a 'green and slum-free' city. At the same time, the governing structures of redevelopment projects are designed to bypass political obstacles posed by the urban poor through delimiting the role of the municipal corporation. The vast slum area of Dharavi in Mumbai was designated as a high-priority 'clustered redevelopment' area and relegated to an entirely separate governance structure, under the auspices of the currently controversial, private architect and developer Mukesh Mehta, a Mumbai-born resident of Long Island, in New York City, who soon became a leading state government's consultant. His 'Vision Mumbai' megaproject was advocated in 2003 as a large policy apparatus to support highly desired urban change (Doshi 2013, 2015), and like in the global North (e.g. mixed communities policy, see Chapter 5), it aimed to bring mixed-income people together from different social strata, implying a form of what Mehta called 'reverse-gentrification'. His plan also asked that the government invite property companies to demolish slums and redevelop the land at a higher density, assuming that Mumbai's very high real estate market prices could be used to cross-subsidize housing for the original slum dwellers in new medium-rise apartment blocks (Harris 2008), something that, as in the global North (see Lees 2014a) did not happen in reality, as we see next.

Newly opened, formerly slum, land markets emerged from multi-scalar, interconnected social and political processes. The slum policy that emerged in the 1990s and that Mehta somehow transformed into a gentrifying goal was not as simple as welfare-oriented interventions but 'a central technology in the social production of both land markets and urban citizenship in the redeveloping city' (Doshi 2013: 848). Over the last fifteen years, there has been a macro-scale operation of urban reshaping at the regional scale in Mumbai, and much of this resembles what is currently going on, in for example, Rio de Janeiro, Brazil, which we looked at in previous chapters, and return to again later in this chapter. The Slum Rehabilitation Scheme (SRS) of the 1990s aimed to remove slums and promote real estate and infrastructure redevelopment at a massive scale. It facilitated slum clearance and leveraged market incentives, while targeting the elite and middle classes who rejected spending public funds to relocate the poor but would accept offering public housing to the 'legitimate poor', land tax revenues for the state, beautification for the upper classes, and redevelopment profits for developers.

The SRS policy worked through a cross-subsidy model based on 'transferable development rights' (TDRs) for builders of 'compensation housing' for the displaced slum dwellers, while granting them the right to build taller buildings somewhere else (Doshi 2013). The latest public-private effort to redevelop Asia's largest slum area is the Dharavi Redevelopment Project (DRP), also designed and promoted by Mukesh Mehta. This project is a higher-income development, comprising luxury housing, office buildings and even a golf course. In 2004, the project received backing from the state to transform Mumbai into a 'slum-free city', but by 2012 construction had not begun for bureaucratic reasons and the resistance of residents (70 per cent of residents had to consent before a block could be demolished). Acceding to Mehta's dream, slum dwellers deemed 'eligible for rehabilitation' were to receive single-room replacement dwellings in mid-rise buildings, free of charge, but they would have to pay taxes and maintenance fees, whereas private developers would receive land as compensation for building these dwellings and infrastructure. Despite the fact that from 1997 onwards Mehta spent significant time building political consensus around the project, and support from the right-wing state government (Ren and Weinstein 2013), thus far his comprehensive project has not been achieved.

The latter case shows the level of influence private interests can have at the local level, where a vacuum of power can be filled by somewhat implausible plans. In Mumbai, informality and ambiguity operate as a system for governing the city and this offers the state the power and flexibility to claim *de facto* tenure and compensation for the urban poor, while also facilitating slum clearances aimed for the production of financial sector development projects. Speculation on land in slums is not new, yet it remains a largely under-studied phenomenon of informal sub-renting and speculative landlordism. Infrastructural improvements in slums cause social displacement of the poorest renters who are superseded by better off lower-middle-class residents offering higher rents or purchasing prices for housing in physically improved settlements; something that Doshi (2015) calls 'marginal gentrification' and can be related to Damaris Rose's account of low income gentrifiers in Montreal in the 1980s (see Rose 1984; Lees, Slater and Wyly 2010). Doshi (2013) also discusses how from the 1990s, thousands of households and entire neighbourhoods deemed 'illegal' – therefore ineligible – were excluded from the resettlement programmes (see similarly Ascensao, 2015, on Lisbon). In addition, ethno-nationalist party politics infiltrated redevelopment practice, and state agencies began to disproportionately target ethno-religious minority neighbourhoods at the top of the list of programmed residential displacements. We believe an alternative label for Dharavi's gentrification could be 'accumulation by differentiated displacement' (Doshi, 2013), a form of not just class-led but also ethnic-led and gender-led residential displacement of minority groups, a result of the extremely polarized power relations that reproduce urban space in Mumbai and other parts of the informal urban world.

MAINLAND CHINA'S 'SLUM' REDEVELOPMENT AND GENTRIFICATION

The word 'slum' is not as useful a label in mainland China when referring to its vast areas of informal, dilapidated and substandard residential settlements, which are increasingly entrenched by newly commercialized hubs of urban redevelopment involving agencies from both the private and the public sectors (but conform to slum characteristics in terms of living conditions). The condensed urbanization in mainland China has not resulted in the extensive emergence of informal squatter settlements that are often witnessed in other countries in the global South, but is thought to be largely the result of the state control of migrants' mobility through a system, which ties welfare benefits to one's place of household registration (hence no benefits from destination places when migrants leave their homes), and also the state ownership of urban lands that turned out to be lucrative resources. There are two types: (a) dilapidation of historic inner-city neighbourhoods, and (b) the rise of urbanized former rural villages that saw a concentration of informal landlordism, resulting in relatively affordable private rental accommodations for migrants at the expense of 'slum-like' conditions of living. As the latter type is discussed in Chapter 7, this section focuses on the case of historic inner-city neighbourhoods.

Many historic inner-city neighbourhoods can be characterized by poor living conditions, including overcrowding and lack of access to basic facilities such as private toilets and bathrooms. Such dilapidation can be attributed to the legacy of the planned economy era. Many dwellings in such neighbourhoods used to be owned and managed by local housing bureaus, an outcome of the socialization during the early days after China's Liberation, as well as various subdivisions and informal construction during the socialist years to accommodate as many families as possible when housing, as part of consumption, saw restricted state expenditure (Zhang 1997). Due to a lack of government funding, even to keep up with basic maintenance, these public rental dwellings eventually fell victim to dilapidation, rendering them exempt from privatization during the housing reform and subsequently becoming target areas when local governments launched state-led redevelopment from the 1990s. The redevelopment process was further accelerated due to the state aspiration to make cities like Beijing and Shanghai look and feel like other world cities, especially as China joined the World Trade Organization and came to win the bids to host major mega-events such as the Beijing Olympic Games and Shanghai World Expo (Broudehoux 2007; Wu 2000, 2002; Shin 2009b, 2012). Many residents were subject to relocation to new housing estates built in suburban districts (particularly during the 1990s) or displacement with cash compensation, when the central state prioritized cash compensation over the provision of relocation, especially after 1998.

Beijing, for instance, has experienced rapid city-wide redevelopment over the past decade in response to state goals that seek to remake the city into a 'world city' by 2050. This is an interesting case for a number of reasons. First, in the state-entrepreneurial goal for enabling property-led redevelopment, the local state aided by a range of municipal institutions basically acts like a *de facto* landlord expelling low-income tenants who possess very few political privileges (Shin 2009b). Second, redevelopment has been guided by a state programme whose very success depends on the participation of real estate capital for financial contributions. Here, real estate capital includes those having both domestic and overseas origins and also that operate in both the state and private sectors. Therefore, large numbers of historic buildings and architectural heritage have been bulldozed to make way for new high-rise towers, and entire locally based ways of life have been erased. Third, low-rise estates, declining workers' villages, factories and warehouses, all built in the socialist period, are now considered to be eyesores for being derelict, slum-like, and inappropriate for the state to maintain, detracting from its global city image, where global architectural styles are used to symbolize modernity and change for the middle classes. While most inner-city neighbourhoods have become subject to wholesale clearance and redevelopment, a small number of historic neighbourhoods deemed of importance in terms of national heritage have been more likely to face gentrification by heritage preservation (see Chapter 5).

FROM THE OLYMPIC EFFECT TO 'FAVELA CHIC': RIO DE JANEIRO

From the 1990s onwards, the Brazilian government has followed an institutional approach to slum eradication that has some similarities with the two previous cases. On the one hand, there are policies of aggressive land formalization via police 'pacification' of slums with displacement pressure on the lowest segments of the original population, who are not eligible for rehousing; on the other hand, there are state-financed policies for rehousing the poor in areas beyond the city's outskirts, as is happening in Beijing and other cities in mainland China with global aspirations. Slums (better known in Brazil as favelas) are numerous in the outer rings of the largest Brazilian cities, but Rio de Janeiro is regarded as one of the few cases where favelas are in fact very central and visible, and therefore the city is currently one of the world's most explicit cases of state-sponsored slum gentrification.

Since the early twentieth century Brazilian cities have been characterized by the implosion of informal settlements, fed by large numbers of newly urbanized local immigrants. In the 1970s and 1980s the Brazilian economy entered into a deindustrialization stage, but favelas remained on the hillsides of Rio de Janeiro, provoking a verticalized form of segregation wherein poverty sits cheek by jowl next to the most affluent residential neighbourhoods and service areas. As Queiroz Ribeiro (2013) claims, since the early 1990s, the whole of Rio de Janeiro's urban system has surrendered to a state-led process of service-oriented economic reshaping to reverse the decline of its post-industrial urban economy. In the 1990s, Rio de Janeiro started mass-scale housing formalization and social domestication in those favelas that at that time were considered the most dangerous ones. Queiroz Ribeiro and dos Santos Junior (2007) observed the breakdown of previous social and productive structures in Rio that served as a support for social reproduction and transformed the relationship between the poor and the rest of urban society. This reconfiguration was paralleled by the weakening of their ties with the labour market and their growing spatial and social segmentation, especially among low-skilled workers. Beyond this economic dimension, the isolation of the urban poor in Brazilian cities also occurred because they shared increasingly fewer services, urban infrastructure and public spaces of employment with the rest of the population; this led to exclusion and stigmatization of the urban poor. But the new post-2000 context brought state-led police 'pacification' and state-supported, class-led, material and symbolic occupation or absorption of cultural capital (or both) from traditional areas where informality of property and labour relations were already in crisis, accompanied by state attempts to penetrate and somehow domesticate the spatialities of urban poverty and informality.

This whole process of urban restructuring apparently creates massive rates of displacement (dos Santos Junior and dos Santos 2014). Cases of real estate pressure and displacement on favelas vary, and it is possible to recognize at least three emblematic cases among many others. First, the explosive and highly mediatized displacement effects from a comprehensive plan to restructure the whole 2016 Olympic zone in eastern Rio de Janeiro; second, the comprehensive redevelopment of the Zona Portuária (old port area) with intervention in the three favelas included within it; third, the much more culturally oriented piecemeal redevelopment of Rocinha and Vidigal favelas, currently infamous for hosting interesting projects of cultural revitalization as well as being deeply 'pacified' by the city's military police (Cummings 2015).

According to Vainer et al. (2013), when Rio de Janeiro won the right to host the 2016 Olympic Games, it was the culmination of a process that began two decades earlier, which chained Rio to a new coalition of power and a 'new city' project. The completion of the Pan American Games in 2007, the Confederations Cup in 2013, the World Cup in 2014 and the Olympics in 2016 provided the concentration of large public investments in the implementation of projects that reconfigured extensive urban spaces and since then have affected the whole city area. A rapid transit bus system increased travel and operations from/to the area of Barra da Tijuca and therefore forced eviction of 20,000 families, as land prices in this hitherto peripheral area of eastern Rio grew on average by 190 per cent from 2009 to 2013 (Vainer et al. 2013). A private company is currently constructing the new infrastructure and services required for the operation of the Olympic Park, and in exchange, after the Olympics, that company is entitled to commercially exploit 75 per cent of the total Olympic area!

Living next to the Olympic area, from 2007 onwards, residents of Vila Autódromo started to be threatened with removal, for several alleged reasons: they were a 'landscape pollution threat', a danger to the safety of athletes housed in the Pan American Village, victims of the high levels of pollution in the pond that borders both the Vila and the Olympic area, subjects incompatible with the implementation of the Olympic Park, occupiers of space needed for the construction of a motorway, and so on. This favela had been colonized in the late 1960s by a community of fishermen who installed temporary dwellings on the edge of the Jacarepagua pond, and remained there ever since, undertaking grassroots self-financed improvements and service provision while they were practically abandoned by the state. Currently, the Vila Autódromo hosts around 450 families, with good standard brick dwellings amid other very precarious ones, and access to water, sanitation and energy supplies. 88 per cent of children and young people study in the immediate areas, while 65 per cent of workers have jobs in the adjacent quarters. Most residents are home-owners and 10.5 per cent pay rent (Vainer et al. 2013). Works for the Olympic area started in 2013 with the aggressive removal of houses and public areas that were regularly used by the community. As in other Brazilian cities, state-sponsored housing programmes offered a rapid solution for those to be displaced if, that is, they accepted relocation to distant peripheral areas, away from the main metropolitan centres of employment. According to Vainer et al. (2013), the Minha Casa Minha Vida (MCMV) state programme for relocation has promoted acceleration of the unprecedented mass removal from Vila Autódromo.

By the end of the 2000s, Rio de Janeiro's 'favela' hillside slums were still virtual no-go zones, controlled by drug lords. Then, in 2011, police began to seize control of dozens of favelas from drug gangs. Rio's slum 'pacification' programme was/is part of a strategy to make the city safe in the run up to the FIFA 2014 World Cup and 2016 Olympics. The Zona Portuária (see Figure 6.2), a three-neighbourhood region that is undergoing massive state-led 'revitalization' and pacification comprehensively named Porto Maravilha, shows how Police Pacification Units (UPP) have been permanently installed in those favelas (as in many others of the city) to vanquish drug trafficking networks and to enforce security, while the national government's Growth Acceleration programme (PAC) includes public investment in several transportation facilities, like cables that connect favelas with the central area, new housing, social services, and new open spaces. In the whole metropolitan region, the 2010 launch of the USD3.9 billion Morar Carioca programme aimed to reurbanize, relocate, or cope with all of approximately 1,000 irregular settlements, at the same time as the Minha Casa Minha Vida programme, played a role in displacement (Cummings 2013). The class-led symbolic appropriation of Zona Portuária and the rest of the favelas in Rio de Janeiro has led to the prospect of their urbanization giving rise to land speculation, newcomers, higher-income residents, and the eventual socio-economic exclusivity of gentrification in these places.

SecoviRio, an organization representing Rio's real estate professionals, estimated that in the 72 hours after police 'took' the first three favelas (in 2011), property prices in those places jumped by 50 per cent, and are still increasing as we write. Dona Marta favela was the first one to be pacified, and by 2010 emblematic, as some of its houses' facades were the canvas for the aesthetic intervention of two Dutch street artists. Shortly after, in Vidigal favela, a luxury boutique hotel with a rooftop pool was built, as middle-class Brazilians and foreigners snapped up properties with wonderful views of the Copacabana and Ipanema beachfront areas from these hitherto forbidden neighbourhoods. These properties are seen as bargains in a city whose real estate prices are among the highest in the Americas (Lees 2014a). A 'favela chic' phenomenon has also emerged, namely the elevation of *favela* culture to global circuits of cultural consumption through its

Figure 6.2: Gentrifying Zona Portuária in central Rio de Janeiro, 2014 (Photograph by Ernesto López-Morales)

associations with cinema, samba and *funk carioca* musical styles (Cummings, 2015). The 'favela chic' name has also been used for a chain of clubs located in relatively expensive areas of London, UK.

Until very recently, in Brazil, there has not been a proper discussion of gentrification, as the use of the term has often been considered to be of little relevance in describing processes of class-led urban conflict over urban space there. However, as the most recent evidence shows, this situation has changed rapidly as Brazil's largest inner city areas, especially in Rio de Janeiro, are being transformed through profit-seeking policies of redevelopment that go hand-in-hand with mega-events.

FROM MEXICO TO SOUTH AFRICA: GENTRIFICATION OF FORMERLY INFORMAL DISTRICTS

So far we have looked at state-led policies of slum gentrification that are transforming the informal or historical nature of large world cities, in powerful emergent national economies. But we have not addressed what happens when, in the absence of comprehensive city-reshaping plans, very spatially bounded, historic, low-income enclaves present hurdles to state-entrepreneurial redevelopment. This is precisely what the subsequent cases of Mexico City, Santiago de Chile, and Cape Town, South Africa illustrate. Mexico City offers a particularly interesting case of the expansion of tertiary activities and conflicts in the low-income, relatively informal Santa Fe area. According to Pradilla (2008), the redevelopment of Santa Fe in Mexico City has largely depended on the city's huge metropolitan state investment in new transportation facilities, and this can be seen as an excellent example of what Rerat and Lees (2011) refer to as 'spatial capital' related to gentrification (see Chapter 3 on this). A 'new logic of urban structuring' is particularly concentrated there, as vast areas are upzoned and penetrated by transport corridors that displace low-income tenants and petty landowners from already formalized, yet highly underserviced, land, and where low-income residents have lived for more than 30 years. According to Delgadillo (forthcoming), the local and federal government undertook evictions of residents who had legitimately and officially occupied part of that land since 1984, when the national government gave away public land lots to local garbage scavengers. Then in the 1990s, some businesses and national and international corporations located next to the area. However, the Tlayacapa site (9. 5 hectares), where 510 houses held a 3,200-inhabitant community, a church, a school and a small market, started to be severely threatened just before 1994, when the Government of the Federal District (metropolitan authority) closed the dump, cleared the site and put in urban services ready for a mega real estate project. The scavengers were targeted for rapid displacement, even though the original residents were legitimately occupying part of the land.

Each family received compensation equivalent to about USD10,000, and scavengers were offered alternative work in a dump on the Eastern fringe of the city, as well as alternative housing there; but they rejected the offer, and were finally evicted in 1994. City government justified the action on the grounds that the 1984 presidential decree was never implemented and the state-led transfer of the land was never properly fulfilled. Therefore, in 1998, the land was 'divested' from the government of the city to a public company that in 1999 sold it to a private university. Two other evictions took place in December 1998 and July 2001, aiming to clear the main access to the land and allow the building of a new, private university's campus (Delgadillo forthcoming). Currently, the Santa Fe area is a concentrated hub of financial and educational services and one of Mexico City's new centralities by 'tertiary corridors' (Pradilla 2008).

Unlike Mexico City, changes in the class composition of residential neighbourhoods in Cape Town are less due to state-led transportation and upzoning policies and more an effect of the societal changes experienced by the country after the end of apartheid, and currently amid a transitional national economy with a growing share of the wine-production/service economic sectors in the national GDP. Neighbourhood change is related to the purchase by real estate business companies of state-subsidized houses originally aimed at low-income beneficiaries, but now desirable to businesses and wealthy families that redirect those properties to their employees. Lemanski (2014) conceptualizes these re-sales as a form of 'hybrid gentrification' as the new residents are not really upper-income groups and are only slightly higher socially than the original vendors, but housing prices have gone up due to the rising demand. Exclusionary displacement is not only experienced by the original beneficiaries of the National Housing Subsidy Scheme who are replaced with slightly higher-income employee-owners, but future low-income residents are also excluded from this booming residential market. This process has similarities with both existing 'downward-raiding' (a traditional process in South Africa that implies the seizure of land and properties by slightly wealthier households) and gentrification. As Lemanski (2014) claims, 84 per cent of the non-original homeowners received property via an employer, while fewer (16 per cent) purchased houses drawing on bank loans. Sometimes, houses rebuilt for newcomers (mostly employees in the wine industry) are larger than previously built, state-subsidized houses. Many manufacturing companies have followed this logic of spatial restructuring for improving their workforce location. Also wealthy families seek accommodation for their domestic workers and gardeners, also contributing to the massive increase of housing prices.

At some point, the largely dominant role performed by non-resident housing buyers confirms that gentrification is a conflict between 'users' with differential economic powers, and not necessarily inhabitants. But formerly, state-beneficiary residents are largely un-, under- and informally employed, while new occupiers are formally and regularly employed, resulting in substantial differences in terms of financial security, employment tenure, and housing upgrades. Although homeowner beneficiaries can ultimately choose not to sell, this 'choice' is constrained by the high costs of homeownership compared to the low incomes required for subsidy eligibility. The 'once-in-a-lifetime' subsidy means that vendors face permanent exclusion from this element of the welfare state, and potentially from homeownership for life. The consequent exclusion is far more permanent than merely being priced out of the area.

EVICTIONS AND THE SETTLERS' INSURGENCY FOR 'THE RIGHT TO STAY PUT'

As we have seen thus far, urban neoliberalism is, on the one hand, strongly pursued by local-state governments, but on the other, it can be contested by increasingly organized and empowered urban social forms of local-level activism that claim a more expanded and particularized right to the city. Successful communities of resistance are those capable of understanding and dealing with gentrification-related structural factors of change, such as the loss of quality of life, direct threats to displacement or displacement pressure, the uneven accumulation of ground rent, the loss of economic and cultural heritage as a form of built environment, the negative environmental effects caused by mega-project construction, and the high levels of displacement generated as effects of the latter. Although some are concerned that the 'right to the city' is a Western European idea (Lees, 2014a), social movements in Latin America, for example, use it along with the term 'gentrification' to fight for their claims (López-Morales, 2013c). If we conceptualize gentrification as defined by capital reinvestment in the built environment accompanying the displacement of existing users, the main tenets of the Right to the City, which emphasizes the taking-back of the power to produce space from the state and capital, may still hold (Lees 2014a; Samara et al. 2013; Shin 2013). But how these tenets are to be realized and the Right to the City put into practice in urban strategies remains subject to various questions and interpretations.

Conflict and resistance need to be contextualized in each locality, critically understood in their temporal and spatial dimensions, and also historicized in the ways in which rights claims have been exercised. Resistance can be seen in several forms, from highly organized and effective activism to silent forms of adaptation and struggles to 'staying put'. For instance, the Right to the City should be seen not just as addressing conflict as a research issue, but also setting out alternative ways of insurgent or conflictual planning, an approach that sees the city from the point of view of its conflicts and therefore offers a rich body of knowledge that can inform and support a new type of political intervention. If top-down planning is seen as an 'efficient' way to prevent, mediate and resolve conflicts, which are seen as dysfunctional, expensive, and threatening to socio-urban cohesion and therefore dangerous to inter-urban competitiveness and entrepreneurial redevelopment agendas, by contrast conflictual planning addresses and unrolls the creative potential of conflict, from which collective subjects emerge to rescue the city as a political arena, as a space in which citizens are confronted, and propelled to arbitrate the development of the metropolis they live in (Queiroz Ribeiro 2013).

An interesting example is Chile. As López-Morales (2013c) and López-Morales and Ocaranza (2012) show, in 2003, the settlers of Pedro Aguirre Cerda (PAC), a district located in the inner city of Santiago, organized and resisted a highly entrepreneurial local-level master plan that, if implemented, would have displaced a high number of dwellers over just a few years. This plan was a managerial-elitist model of policy delivery, seeking to attract property-led, large-scale renewal, with a strong intervention in local land economics, and therefore producing increased exchange values and thus spatial exclusion. In PAC, more than 50 per cent of residents share residences despite a very high rate of approximately 80 per cent owner occupation (already individually entitled during the Pinochet dictatorship). If PAC's new plan had been legally approved, it would have cleansed several traditional low-income areas to attract middle- and high-rise renewal, and thus significantly altered the social composition of the PAC district. However, PAC's local community realized that even by accepting the changes in the urban and building codes contained in the new master plan, larger rent gaps did not necessarily mean higher land prices to be paid to traditional inner-city, low-income, owner-residents as a form of capitalized ground rent (as we saw in Chapter 3). In fact, it was the opposite. Most households in Santiago's inner city are multi-nuclear, making the actual low capitalized ground rent clearly insufficient in use value terms for those numerous, agglomerated households to find replacement accommodation elsewhere in a highly segregated and expanding city of 65,000 hectares, hence large-scale displacement. Although the main goal of local-level activists was the redrafting of the PAC district master plan, their opposition was at the same time a contestation of state-led class-power exerted over the remaking of a working-class community, of the unaccountable institutions of governance that characterize the Chilean planning system, and of its top-down practices. PAC leaders mobilized and disseminated core technical knowledge at the grassroots level, achieving a complete reshaping of the whole process of masterplan implementation, and made the state redraft its original plan into a newer version of the plan that did not present threats of evictions. Key problems in the district are still unresolved, like the low quality of many of the existing units, but no evictions or displacement pressures have been seen since these events happened, and this is seen as a tremendous success by local residents and other low-income communities in the city (López-Morales, 2010, 2011, 2013a).

166 SLUM GENTRIFICATION

Figure 6.3: People of La Victoria doing participatory planning, 2011 (Photograph courtesy of Matías Ocaranza)

In similar vein, Cabannes et al. (2010) published a report on forced evictions around the world, offering detailed information on strategies and practices deployed by urban dwellers to stay put and oppose redevelopment motivated by private economic agendas. Even if not all of these cases can be deemed gentrification and some are urban and some rural, the forms of resistance shown in this report and the ways people face evictions are useful:

1. Public protest and direct confrontation, including street demonstrations and marches, picketing and blocking off streets (in Buenos Aires and Durban this practice is quite common), draping buildings

that are to be demolished with slogans claiming rights and constitutional guarantees (China), etc.
2. Legal battles, as most organizations facing evictions take an extremely 'legalist' and rights-based approach, and carry out their battles using legal procedures to the maximum extent possible, as the Chilean case shows.
3. Negotiations while resisting, and the claim of 'in situ relocation', which means relocation into neighbourhoods that are within walking distance of the place where the people under threat of eviction are living.
4. Internal mobilization, federated mobilization (beyond the neighbourhood, including other organized settlers), and a search for international solidarity and international networks, including the mobilization of the local and international media.

CONCLUSIONS

The work of Cabannes et al. (2010) above teaches us how evictions are being resisted around the world. However, we think their work also represents a traditional perspective on slum conflicts, a narrative that understands eviction and displacement as very present threats, but where dialectical reflection on class-motivated disputes over space (usually embedded in policies and urban renewal master plans) are not strongly evident in their analysis. Also the changing nature of cities in the global South needs to be recognized, not just for industrial expansion or modernization but as expanding spaces of landed capitalism, and where national states are increasingly relinquishing their role as redistributive agents of public goods, and are now more deeply engaged with cycles of landed capital reproduction and accumulation. Slum evictions are not just to be fought against; they are complex chains of urban change and property-led revitalization that need to be understood at several scales, amid the rapidly changing contexts of contemporary and globalized urbanization. Also, the most positive, modernizing perspectives on slum redevelopment that transpire in the most progressive NGOs or transnational agencies, such as the UN Habitat's political goals for redevelopment, not only miss the same points as raised by those who simply fight evictions, but also too often glorify social mixing and underestimate the threats of displacement.

Both the most negative and positive views on slum redevelopment do not usually see the whole picture of contemporary urban change, and this is precisely what the gentrification lens can offer – something new to understanding slum redevelopment in a comparative fashion, in a rapidly urbanizing world. Eric Clark's (2005) definition of gentrification allows us to see how most urban societies under capitalism in the global South (but also North) are continuously being polarized, while the right to the city and the right to stay put is continuously obliterated. We should also investigate how the injection of capital investment (increasingly coming from global capital) into some areas is an effective threat to socio-spatial stability in urbanizing societies, and the ways these threats work. The cases discussed in this chapter show that property-led (re)investment in spatially constrained slum areas, the visibly changing urban landscapes of the global South, the social 'upgrading' of locale by incoming (or the interests of) higher-income groups, and the direct or indirect displacement (or threats of displacement) of the lowest income groups in society as a whole, are four factors usually occurring at the same time in many places in the 'informal' urban world. The role of the state and international donors and NGOs cannot be underestimated as part of the politico-economic repertoire of current slum gentrification. As Desai and Loftus (2013: 790) rightly claim:

> [w]ithout a clearer understanding of the circulation of capital through land in slum areas there is a serious risk that benevolent investments in

infrastructure will merely strengthen powerful stakeholders operating within and across slum areas, while increasing the already precarious nature of slum dwellers' lives.

Slum eradication for the expansion of middle-class consumption and the realization of the state's entrepreneurial spatial agenda, that is 'slum gentrification', is not only a goal to open up new spaces for real estate or service-oriented investment, but also a way to erase the past and, when that is not possible or strategic, to commodify its physical fragments into a new spatial commodity that is trendy and saleable. Currently, slum gentrification in Beijing's hutongs or Shanghai's lilongs, or in Rio's favelas, very much resembles the classic gentrifier's quest for urban excitement and difference (see Lees, Slater and Wyly, 2008, chapter 3). What is different now, however, is the scale of the negative outcomes experienced by the low-income slum residents, for the scale is a hundred times larger, an issue that we turn to in the subsequent chapter. These outcomes may well be transforming the whole urban socio-economic system of Southern metropolises, in a more permanent way than was done in the gentrification of the 1960s and 1970s in the global North.

Often, slums are mistakenly thought to be excluded from urban real estate markets, not contributing to the urban economy, and this has traditionally led authors like Hernando de Soto (2000) to advocate the 'inclusion' of slums through title formalization so that the hidden economic potential of slums can be fully developed in the formal market and original residents can become agents with economic assets. But once land and properties in slums are regularized through individual titling, as we have seen in this chapter, land and housing prices generally increase, forcing lower-income residents to sell and move to cheaper unserviced areas on the periphery of cities. Therefore Lemanski's (2014) narrative of 'hybrid gentrification' or López-Morales' (2011, 2013a,c) 'accumulation by ground rent dispossession' seem appropriate terms to refer to cases where the real purchasing power is not in the hands of incoming residents but in the hands of real estate investors and the state's concerted efforts to implement 'state-led' gentrification or 'pacification' to invite real estate capital into slum areas. We also think these concepts are not exclusively relevant to Southern cities but serve as models for theorizing across Northern and Southern cities more broadly.

7 Mega-Gentrification and Displacement

As the world witnesses deepening urbanization taking place in a highly uneven manner across and within regions and continents, the pace and scale of urban development projects has also ascended to an unprecedented scale. Contemporary cities in both the global North and South increasingly promote mega development projects. Dubai, Abu Dhabi and Bahrain, for instance, have been implementing new forms of urban development – most of which has been focused on large-scale infrastructure and real estate projects marketed at transnational capital and the global rich. Cities in mainland China have also been replacing low-rise workers' compounds and run-down industrial complexes with commercial real estate projects, each project directly affecting thousands of existing residents. Cities like Hong Kong have seen the demolition of individual residential high-rise towers in central districts to be replaced by more lucrative commercial or residential buildings, again affecting hundreds and thousands of households. To what extent can we refer to these projects as gentrification? To what extent do state-led developmental projects constitute gentrification? How do different factions of capital operate to extract exchange value at the expense of existing owners and renters' use value as well as their right to stay put?

In his afterword to the 2008 edition of *Uneven Development*, the late Neil Smith reiterated his earlier claim, stating that in the period of contemporary neoliberal globalization, '[a]t the urban scale, the gentrification which in the early 1980s was still an emergent phenomenon, has become a global urban strategy' (Smith 2008: 263). Cities of the global South are regarded as having faced the brunt of the latest wave of 'global' gentrification, as mega-scale gentrification accompanied by mega-scale displacement (Lees 2012: 164). Examples from around the world testify to the severity of the mega-displacements resulting from the implementation of a particular set of urban policies and mega-projects, which aim to transform existing urban spaces within a highly compressed time scale. Mainland China, among other countries, seems to stand out in the discussion of mega-gentrification. In the case of Beijing, its official estimates indicate that the city's redevelopment projects throughout the 1990s affected about half a million residents or 160,900 households, resulting in the permanent displacement of more than two-thirds of the affected (Fang and Zhang 2003). Between 2000 and 2008, again based on official estimates, it was further reported by a Geneva-based international NGO that nearly 1.5 million residents were expected to incur the brunt of redevelopment projects in Beijing, a process accelerated in part by the city's preparation for the 2008 Summer Olympic Games (COHRE 2007). Official figures mostly include the number of permanent residents without taking into account migrant workers and their families, and therefore, the actual impact on urbanites in Beijing would well exceed one's imagination. The situation was similar in Shanghai too, where the municipal government's ambitious redevelopment projects to transform the city affected about 746,000 households between 1995 and 2005 (see Iossifova 2009), and half a million households between 2003 and 2010 (Shanghai Statistical Bureau 2011 cited in Shin 2012). But mega-displacement experiences comparable to those of China's cities can also be found elsewhere, especially in its neighbouring country, South Korea. New waves of commercial redevelopment projects in Seoul in the 1980s, partly accelerated by municipal preparation for the 1988 Summer Olympic Games, reportedly displaced 720,000 people during the five years between 1983 and 1988 (see ACHR 1989: 91). This meant that nearly one in every ten people in Seoul at the time were subject to displacement.

Gentrification in its classic form, as first discussed by Ruth Glass (1964a), incurred a reasonably long-term process of neighbourhood change, involving dwelling-by-dwelling rehabilitation to convert working-class accommodations into those for a more affluent class. The pace of classic gentrification would often be slow enough for local residents not to notice the severity of the changes until the transformation became irreversible. However, the subsequent waves of gentrification (see Hackworth and Smith 2001) have been much more visible and vicious, involving a series of residential blocks that accommodate hundreds, if not thousands, of residents who become subject to mass displacement to give way to wholesale clearance and reconstruction of upmarket buildings and facilities that lie beyond the reach of the original residents. The massive scale of redevelopment in contemporary cities, and hence the domination of new-build gentrification, suggests a whole set of different rules and norms for the operation of the state, its apparatus, including planning authorities and law enforcement agencies, and real estate capital; that is, a different set of rules and logics of accumulation, which is qualitatively different from the gentrification Ruth Glass discussed.

In this chapter, we aim to contribute to extending understanding of contemporary gentrification that has become not only global in accordance with extended or planetary urbanization (Brenner and Schmid 2014; Merrifield 2013a; see Chapter 2) but also massive in scale to produce mega-gentrification, which elevates the effects of domicide to a whole new terrain for the displaced (Porteous and Smith 2001; Shao 2013). Our enquiry into global mega-gentrification calls for research on the changing nature of the state and capital as well as state-society relations. In doing so, the chapter pays special attention to the contextual nature of gentrification in each society, for gentrification 'is a product of particular historical, contextual and temporal forces' (Lees 2012: 165). Contextualizing gentrification in a society also means that we refrain from isolating gentrification in a country as being unique and incomparable with gentrification elsewhere. While mega-gentrification in the global South appears to be more prominent and brutal, we are aware of equally brutal and vicious processes of domicide as a result of gentrification in the global North.

MEGA-DISPLACEMENT AS A PRE-REQUISITE FOR URBANIZATION IN THE GLOBAL SOUTH?

We begin by enquiring into the rise of mega-gentrification in the global South. This enquiry requires us to make a further distinction between displacement-led development and gentrification, for many countries in the global South are undergoing massive restructuring of their national territories, which involve spatial restructuring to accommodate various land enclosures and clearance efforts to accommodate developmental projects such as railways, dams, power plants, factories and ports, as well as other essential urban facilities. Informal settlements often located in the path of such developmental projects are subject to clearance, resulting in mega-displacement known as 'development-led displacement' (DID). It is necessary to separate these processes of DID from the process of gentrification, although very often, the promotion of mixed-use projects suggests that DID and gentrification may occur simultaneously or sequentially in some instances.

In discussing spatial fix as a solution to overcoming the overaccumulation crisis in industrial production, David Harvey (1978) highlighted the ways in which capital switching took place to channel surplus capital into the built environment, especially into expanding fixed assets such as infrastructure (motorways, rail, power plants and so on) and also the real estate sector. This has become recognized as a prominent feature that highlights the several rounds of economic crises that characterized economies such as Britain and the United States from the 1970s. The experiences of the global South, in particular countries such as mainland China, have provided an opportunity to further refine theories about circuits of accumulation and capital switching. Rather than the secondary circuit of the built environment rising above the primary circuit of industrial production, the two circuits are often mutually supportive in countries that lack production bases as well as infrastructure and facilities (see Shin 2014a: 510–12).

The first dimension of the channelling of capital into investing in fixed assets produces displacement, which can be conceptualized as displacement by development or development-induced displacement (DID) that largely takes the form of forced eviction and resettlement (Mehta 2009; Satiroglu and Choi 2015). Given the large-scale nature of most development (especially infrastructure) projects, permanent displacement takes place often at the scale of hundreds of thousands of people per project. For instance, 192 dam projects around the world, which had entered into the construction phase between 1986 and 1993, reportedly produced the displacement of four million people annually (Bartolome et al. 2000: 4). In Manila, the Philippines, its metropolitan rail project accompanied large-scale land assembly, resulting in the clearance of informal settlements along the routes. In total, 35,000 households faced displacement between 2003 and 2010 (see Choi 2014). In Beijing, as part of its preparation for the 2008 Summer Olympic Games, the municipality aimed to carry out so-called environmental improvement projects, which would have affected about 370,500 people in migrant-concentrated 'urbanized villages' (Shin 2009c: 133–5). The list could go on indefinitely. The bottom line is that mega-displacement has become synonymous with development in the process of capitalist accumulation. DID shares a number of features with gentrification. In particular, those affected by DID may go through similar experiences of displacement as experienced in times of gentrification, given their eviction from homes and neighbourhoods. Narae Choi (2014: 9) thus notes: 'The process and outcome of such mass displacement and relocation and, most importantly, the uprooting experience felt by displaced residents were almost identical to those of gentrification-induced displacement, particularly in terms of the issues around compensation housing and the challenges in a new site, which often result in the return of the relocated'.

DID does not always lead to gentrification. Bartolome et al. (2000: 7) point out that DID has four major characteristics: (a) displacement results from a particular model of development that does not give consideration to minimizing social and environmental costs; (b) most displacement involves involuntary displacement of people who are prevented from making meaningful participation to voice their views; (c) displacement often incurs a long period of refusal of development opportunities and traumatic forced relocation due to insufficient warning; (d) the scale of displacement is often underestimated. Obviously these characteristics echo those of displacement in the gentrification literature, but ultimately, DID differs from gentrification, given the centrality of infrastructure projects. Mega-displacement can be a daily experience of both urban and rural inhabitants in the global South, which occurs as a result of development projects that are supposedly to benefit the entire population, however dubiously public interests are defined in reality. An illustrative example would be the construction of metro systems in China's major cities such as Beijing, Shanghai and Guangzhou, which incurs the displacement of hundreds of thousands of urbanites but benefits entire populations, including low-skilled migrant, by providing heavily subsidized fares that have hardly increased for years.

INFRASTRUCTURE DEVELOPMENT AND GENTRIFICATION

However, it is also necessary to discuss how DID indirectly produces gentrification under particular circumstances by means of creating rent gaps through upping potential ground rents by the installation of infrastructure and urban amenities such as parks and new water fronts. In the earlier cited example of Manila's metropolitan rail project, it was hinted that '[t]he removal of informal settlements [as part of land clearance for rail construction] has the secondary effect of formalizing the land use and access to infrastructures in the locality as well as of physical "beautification", thereby opening doors to formal development/redevelopment' (Choi 2014: 12). This creates greater vulnerability for the existing urban poor as they are placed under displacement pressure due to ensuing speculation and increased living costs. In the case of Taipei, the construction of Da-an Forest Park, dubbed as Taipei's version of Central Park in Manhattan, led to a surge in land prices and the proliferation of lucrative real estate projects in areas adjacent to the park (Huang 2015; Jou et al. 2014).

A more recent technique for boosting development potential frequently used is transit-oriented development (TOD). Hong Kong in particular has been experimenting with this to ensure that its state-owned lands are put into the more lucrative uses of development when they are assembled for expanding its metro network. By means of implementing a 'rail + property' development programme, the development rights in the immediate surroundings, as well as the space above each station, are sold to real estate developers for mixed-use development (see Cervero and Murakami 2009). While the revenues generated from such a sale of rights have contributed to the Hong Kong MTR Corporation's recovery of metro construction costs, upmarket residential and commercial development around each station precinct has also resulted in a price premium for the residential properties on the development site. This creates further opportunities to put land uses in adjacent areas into a higher and better use, adding more displacement pressure (also see Rerat and Lees 2011 on spatial capital around Swiss railway stations).

Creating new urban amenities through public investment may also trigger gentrification by inducing the increase in land values in surrounding areas. In this case, development projects for installing such amenities may be regarded as creating a beachhead for subsequent gentrification. This is exemplified by an urban project in Seoul, which was to restore an inner-city stream that was covered nearly half a century ago. Known as the Cheonggye Stream Restoration Project, it is a project that resulted in the provision of a 5.8-kilometre open-air stream with footpaths as a linear park that traverses the historic centre of Seoul (Lim et al. 2013). It is known for its green nature and gained world-wide attention in recent years (e.g. Revkin 2009). While the project itself was publicly announced in July 2002 and saw its completion in October 2005, Lim et al. (2013) examined the changes in land uses between 2000 and 2011 in historic central business district areas divided by a section of the restored stream. One of their key findings was confirmation that land prices and rents increased after restoration, which resulted in 'accelerating land use change as land owners seek to maximize profits by attracting more affluent users who value the newly created urban open space' (Lim et al. 2013: 199). Some refer to the leading role of public entrepreneurship (Seo and Chung 2012), although in reality the project has seen the advancement of urban elites' political aspirations, as well as spatial restructuring to open up further avenues of urban capital accumulation in the historic central business district, which was redeveloped two-to-three decades ago and was awaiting another round of revalorization. The result has been the advancement of 'commercial gentrification', displacing small businesses and industrial uses that used to take advantage of low rents before the restoration project, giving way to more

Figure 7.1: Restored Cheonggye Stream in Seoul, 2011 (Photograph by Hyun Bang Shin)

affluent users and the intensification of commercial functions for high-end users (Lim et al. 2013).

MEGA-GENTRIFICATION, THE STATE AND URBAN ACCUMULATION IN THE GLOBAL SOUTH

We have seen in the previous section how DID occurs on a large scale, especially when infrastructure projects take place in concentrated urban areas with a high density of residential populations in the global South. Increasingly, these residential areas, both formal and informal, are subject to speculation (c. f. Desai and Loftus 2013; Goldman 2011; Shin 2014a). This relates to a dimension of capital switching into the built environment corresponding to the consumption fund in David Harvey's schema. This section discusses how mega-gentrification has been occurring, and what made such mega-gentrification possible. Gentrifying a crowded inner-city neighbourhood area in its entirety, for instance, usually goes beyond the scope of an individual developer or pioneering gentrifiers. Such a task requires the intervention of local and central states, often entrepreneurial, to make it possible. In this regard, we start by examining the intervention of the state in gentrification, which has been more strongly pronounced in recent years. In this section, we introduce several cases in order to discuss how conflictual and/or supportive interactions between the state, capital and citizens produce a particular typology of gentrification and displacement. Obviously these cases (and those sub-cases under each category of the state) are not meant to be exhaustive but suggestive of further investigation.

Mega-gentrification under East Asian developmental states

East Asia as a region has been known for the developmental orientation of its national states, which ensured their legitimacy by achieving economic security for its general population despite its undemocratic and authoritarian nature of governance in times of condensed industrialization (Woo-Cumings 1999). Having benefited from preferential access to the US market in return for their support for the US-led capitalist bloc during the Cold War period (Glassman and Choi 2014), East Asian developmental states were investing heavily in fixed assets to build up their infrastructural strength to pursue, initially labour-intensive manufacturing industries, then moving up the global value chains through technological innovation and exploiting neighboring emerging economies with cheaper labour costs (Arrighi 2009; Kim 2013; Yeung 2009). In the housing sector, socio-political relations determined the ways in which the national state intervened in housing provision for their national population. Bae-Gyoon Park's (1998) comparative study of public housing provision in Singapore and South Korea is illustrative in this regard. In Singapore, public housing was extensively supplied in part to secure social stability for the social reproduction of labour for its nascent industrial production, and in part for the state's acquisition of legitimacy by being able to provide for the immediate needs of the middle and working classes. In contrast, South Korea was seeing the domination of a growth alliance between the state and large businesses (commonly known as *Chaebols*), as the state redirected available resources towards helping businesses grow and secure their share in the world (the US in particular) market. While the state helped build the real estate industry and strengthen the housing sector through the activities of its national housing corporation, these were mostly to strengthen the commercial housing market based on freehold ownership so that public housing stocks were no longer part of the state's responsibility (Park 1998). Families were to cater for their own housing needs, and during the earlier period of condensed industrialization and urbanization in the 1960s and 1970s, the growth of squatter and substandard settlements was inevitable, as the urban poor were finding it difficult to gain a foothold in the formal housing market.

It is in this context that the Seoul municipal government in South Korea embarked on a city-wide programme known as the Joint Redevelopment Programme to transform its prevailing substandard settlements into commercial housing districts for the middle classes from the early 1980s. The programme provided a platform for a coalition between property owners and developers to work together to initiate commercial redevelopment. The city at the time was also in desperate need of beautifying the city-scape in preparation for the 1988 Summer Olympic Games. The municipal government resorted to a strategy of wholesale clearance of existing structures and reconstruction of upscale commercial flats to the highest density permitted by planning regulations. The Joint Redevelopment Programme (JPR) was technically a programme led by an association of property owners in substandard settlements, who would choose a developer or a consortium of developers as a redevelopment partner for the provision of finance, project management, construction and sales of completed flats in the new housing market. Legalization of land tenure was part of the programme, if property owners possessed illegal dwellings on state-owned lands, a common feature that characterized many of the surviving substandard settlements at the time. While property owners and developers were to be in the driving seat of redevelopment projects, the local state (the municipal government and other state apparatus including the police force) were acting more as a facilitator, actively intervening through its use of planning and administrative powers. While the JRP had the nominal aim of turning poor owner-occupiers in substandard settlements into homeowners of new modern commercial flats after redevelopment, in practice, upmarket new flats were often beyond the reach of poor owner-occupiers despite the preferential discounted prices of flats for the members of property-owners' associations. Speculative interests often came into these redevelopment areas, and more affluent buyers often purchased the property rights (often referred to as *ddakji* in Korean or tickets) to replace existing poorer owners who attempted to acquire marginal gains (that is, a portion of potential ground rents) by selling their rights. Successive transactions often took place motivated by speculative interests.

Based on government-supplied data, Shin and Kim (2015) suggest that out of 211 project sites subject to the redevelopment programme completed between the early 1980s and 2010, the average size of dwellings per project site turned out to be 379 units. Thus, it was not uncommon to see thousands of people evicted from a neighbourhood that had been an affordable place of residence. Authoritarian states intervened to guarantee the smooth operation of profiteering activities launched by a faction of property owners and developers. Brutal evictions were common, especially during the 1980s and early 1990s, often exercised by thugs hired by developers and helped by the police who coalesced with developers. Here, the case of Sanggyedong redevelopment is an exemplary case of how the state and developers pushed forward with mega-gentrification by implementing a joined-up effort to undermine the housing rights of the urban poor.

Case study: Sanggyedong redevelopment in the 1980s

The redevelopment of Sanggyedong resulted in the displacement of 1,528 households, which included 947 squatter owner-households without land titles and 581 tenant households (Kim 1998). It was designated as a redevelopment district on 20 April 1985. The metro station opened in January of the same year. The size of the Sanggyedong project site reached 43,620 square metres, and the area was a substandard settlement located in the then northern suburban area of Seoul. According to the local district government, the substandard settlement was standing on lands mostly owned by the central state (and a small share of municipal land), hence rendering the neighbourhood as informal and illegal. As noted earlier, most original owners sold their property rights to speculators: according to a study, it turned out that '[o]f all the houses that had switched ownership, 18 per cent had changed owners more than five times' (Kim 1998: 214), and eventually less than 10 per cent of the original squatter owners were rehoused in the new redeveloped estate on the site (ibid.). On the other hand, tenants were facing severe disadvantages and permanent displacement, which took place in the most violent way that one could envisage. The following excerpt from a report put together by the Asian Coalition for Housing Rights vividly captures the disastrous experiences of those residents, most of whom had been tenants, when they were resisting eviction:

> There were around 1,100 houses on a four hectare site. When local residents organized to resist this redevelopment (and their eviction), they were subjected to a series of violent attacks; between June 26th 1986 and April 14th, 1987, they were attacked 18 times. Around 400 people (about half of them young people and children) resisted eviction. Most of these attacks involved several hundred riot police and several hundred men hired by the construction company to intimidate and assault the residents. Many of the residents suffered serious injuries as a result of these attacks – including grandparents and babies. The cost of treating those injured totaled some US$15,000 during this period . . .
>
> On April 14th, 1987, a force of 3,500 people and 77 trucks moved into the area – against 380 citizens. The belongings of the residents were loaded onto the trucks and driven away. The tent headquarters was demolished and the residents carried off the area. Wide trenches were dug at the entrances to the area and large barricades erected to prevent re-entry. Riot police and guards hired by the construction company remained to guard the site. (ACHR 1989: 92)

The case of the Sanggyedong eviction received greater attention than other eviction cases in the 1980s, probably because of the tenant struggle's direct connection with the 1988 Summer Olympic Games, showcasing the brutality the authoritarian regime imposed upon evictees. As Davis (2011: 592) states, '[t]he Sanggyedong group debacle raises the question of how much eviction and development was taking place, where the human impact was practically invisible'. For us, it is also a testimony of the brutality associated with mega-gentrification carried out under a developmental authoritarian state, whose action is nevertheless supported in part by speculative property interests including members of property owners in redevelopment project sites and other speculative buyers who replaced poorer owner-occupiers therein. These evictions due to gentrification took place a whole year before the revanchist evictions in Tompkins Square Park, in Manhattan, New York City (see Smith 1996)!

Mega-gentrification in neo-authoritarian socialist states

Large-scale displacement in mainland China is not new. Having experienced accelerated urbanization since the implementation of economic reform and Open Door policies, the central and local states have guided a huge amount of investment in fixed assets to expand the country's economic growth (for a summary of China's urbanization process, see Shin 2015). This process has entailed expropriation of village lands and the conversion of farmlands under the ownership of village collectives into construction lands to be owned and managed by the state. As Carolyn Cartier points out, '[w]ith the state and village collectives as sole landowners, the crux of the problem is that local governments and officials appropriate and use land as a negotiating tool to attract investment and marshal authoritarian power to put down protests against land appropriation' (Cartier 2011: 1117).

The power to expropriate land for development extends to urban governments, who are continuously seeking land resources to maximize their revenue generation on the one hand, and on the other, to transform the urban space under their jurisdiction into modernized space to realize world-city aspirations and cater for the consumption needs of the emerging middle classes. While many lands on the periphery are dedicated to further industrialization (e.g. special economic zones for export-processing and manufacturing) and technological innovation (e.g. science parks) through offering cheap lands to attract investment, lands in existing urban areas are increasingly subject to high fees upon sales of land use rights to developers (in both the state and the private sectors) who are to initiate various real estate projects. The land use premiums collected from the use right sales constitutes the major part of extra-budgetary revenues for local government finance. The growing importance of land revenues in financing urban development in Chinese cities has given rise to the conceptualization by Chinese scholars of 'land-based accumulation' (see Hsing 2010 in particular).

The production of mega-gentrification in mainland China on the immense scale outlined at the outset of this chapter is an accumulated outcome of numerous individual projects that have been implemented across time and space. Obviously, all these projects are not identical in terms of their institutional backgrounds, history of project sites and social consequences. Nevertheless, given the context of China's real estate investment having become a major contributor to the country's fixed asset formation and thus economic development and urbanization, it can be reasonably concluded that mega-gentrification is closely knitted with the urban development agenda by the Party State. It also involves in particular local states at the municipal scale, which have displayed increasingly entrepreneurial characteristics (Duckett 1998; Shin 2009b; Wang 2011), pursuing profiteering opportunities at the expense of local inhabitants' rights to the city and to stay put (Iossifova 2009; Shin 2013; He 2012). Two case studies of top-down redevelopment are illustrative in this regard, exhibiting the highly interventionist, entrepreneurial and authoritarian nature of the local state. These case studies are largely taken from Cheng (2012).

Case studies: Top-down redevelopment in Xi'an and Foshan

Two case studies discussed here exhibit how the local states in mainland China have been actively promoting property-led redevelopment in order to address the accumulation agenda, as well as place promotion that meets the state elite's aspiration to build a modern city that builds upon commodification of real estate properties and breaks away from the practices of China's planned economy era. The two specific cases refer to the redevelopment of a former industrial and residential compound in Xi'an, and the redevelopment of an extensive heritage neighbourhood in Foshan.

In the case of Xi'an, a regional centre in northwestern China, the site concerned was a mixed-use compound in an inner-city area of 5.3 square kilometres, known as Fangzhicheng that accommodated about 160,000 people. The compound was built by the central government shortly after China's Liberation, and used to provide jobs and living amenities for inhabitants whose livelihoods used to depend heavily on five state-owned textile mills and other supporting facilities and institutions. When the economic reform started, the place lost its competition with those textile industries in the coastal region, and experienced reduced operation or closure, laying off a number of workers whose livelihoods deteriorated as a result. When the local government acquired control over the compound from the central government in 2008, it subsequently embarked on the redevelopment of the area into a mixed-use area to accommodate residential, business and industrial functions. A government agency called the Fangzhicheng Integrated Development Office was formed to oversee the whole process, delegated with full powers to administer the operation and approve plans. The agency also established a tight working relationship with developers interested in the project, attracted by the locational advantage of the project site due to its proximity to a new metro station and the high demand for housing in adjacent areas.

In the case of Foshan in Guangdong province, the site of redevelopment involved a residential neighbourhood known as Dongguanli that was well-known for its architectural heritage and designated as a national historic preservation site in 2001. The neighbourhood accommodated about 30,000 residents or 9,635 households in its 0.64 million square metres in addition to a number of small businesses. Its redevelopment commenced in early 2007, directly led by the municipal government of Foshan, which learned extensively from the experience of Shanghai's Xintiandi redevelopment (Yang and Chang 2007; Wai 2006) and made sure that the same developer (Hong Kong Shui On Group) for the Xintiandi project would develop Dongguanli. Despite having heritage preservation as the key objective, the actual project, as demonstrated by the first phase completed in 2011, incurred the near complete demolition of existing structures on the site (as was the case with Xintiandi) and the reconstruction of new structures with antique styles, reminiscent of the 'fake-over' process in Beijing's Qianmen redevelopment (*People's Daily* 2009; see also Chen 2011) (see also the section on historic preservation in Chapter 4).

In both cases, the local state made heavy-handed intervention in the redevelopment process. In Xi'an, compensation offers to local residents living in decades-old self-built units were against the residents' original expectations, and far from adequate to purchase redeveloped units even at discounted prices, as many households were financially stricken. While local residents launched a number of individual and collective actions, formal and informal, to express their discontent, the government agency in charge of the redevelopment of Fangzhicheng took various measures to make sure their discontent went unnoticed. These measures included: (a) deletion of online petitions from web sites and prohibition of media reports on the case; (b) rejection of residents' property rights claims on their self-built units despite their recognition by a state ministry in 1956 upon their construction; (c) backroom negotiation over compensation standards; (d) explicit propaganda activities in the neighbourhood to compel residents to agree to demolition; (e) threats to cut electricity and water supply if residents refused to sign compensation agreements; (f) coopting laid-off workers through their employers by offering priority arrangement for re-employment and additional options of living in low-rent flats on site if unable to afford to purchase redevelopment flats. In Foshan, the local government was more actively involved as it carried out the demolition works itself rather than outsourcing them in order to ensure timely assembly of land for developers. Residents in Dongguanli were subject to off-site relocation as developers aimed to maximize their gains from commercial redevelopment of the site, which targeted the rich. While the residents were offered the highest compensation standards in the city, incidents of discontent still broke out due to dissatisfaction with the compensation measures that did not permit residents' return to the site after redevelopment. Their online protests were banned, and media reports were also prohibited so that the voices of frustrated residents would go unheard. Local taxation officers often disrupted the day-to-day running of small businesses by exercising frequent checks on their accounting records, a means to coerce them to sign compensation agreements and vacate their premises. Public servants living in the redevelopment site were also subject to coercion by their employers, who urged them to sign the compensation agreements so that demolition could proceed. Between 2008 and 2011, dwellings were demolished one-by-one by the local government's exercise of forced demolition.

The actions of local governments in both Xi'an and Foshan were in close cooperation with real estate developers to make sure their redevelopment sites were to be ready for the injection of real estate capital. Local residents' rights to their neighbourhoods and to the city were dispossessed in the process, reminiscent of David Harvey's 'accumulation by dispossession' (Harvey 2005, 2010b). In the context of speculative urbanization in mainland China where real estate investment has become a pillar of economic development, both the Fangzhicheng and Dongguanli projects could be regarded as an effective means to release state land so that the space could be converted into a commodified space for extracting exchange value (Weber 2002; see also Shin 2014a), supported by the actions of the local state acting as a *de facto* landlord (Shin 2009b).

HIDDEN DISPLACEMENT IN THE GLOBAL SOUTH

Large-scale mega-gentrifications are often referred to by critical scholars and advocacy groups as phenomena that produce a qualitatively different experience of displacement from those in the global North. Referring to the reports by COHRE (2006) on forced evictions, Elvin Wyly and his colleagues wrote:

> The only circumstances where there is *not* an obsessive concern with the uncertainty of numerical estimates are those places in the global South where the scale of dispossession is undeniable: an estimated 2 million people were displaced by forced evictions across Africa between 2003 and 2006, and 3.4 million were displaced in Asia – not counting the million-plus pushed out to make way for Beijing's hosting of the 2008 Olympic Games. (Wyly et al. 2010: 2603; original emphasis)

However, the scale of displacement as reported above is mostly based on counting last-remaining residents who were occupying property at the time of demolition and/or forced eviction and directly affected by it. A more nuanced understanding of what displacement actually constitutes tells us that such a method grossly underestimates the actual scale of displacement.

In his thought-provoking piece on the processes of gentrification, Peter Marcuse (1985a) rejects a simple definition of displacement as equating with only 'direct' displacement that affects the last-remaining residents. In addition to 'last-resident displacement', Marcuse further adds 'chain displacement', 'exclusionary displacement', and 'displacement pressure' (see Marcuse 1985a: 205–8). Chain displacement occurs when households who occupied a unit previously, before last-remaining residents, have also been subject to displacement. Exclusionary displacement results when previously affordable dwellings and neighbourhoods become gentrified or become no longer available (due to various reasons e.g. law and planning regulations) for those households who could afford to live there before changes. Finally, displacement pressure is what may be felt by those households who are not immediately affected by displacement but whose experience of displacement may become a reality in time to come. Such displacement pressure may be generated due to on-going changes in the neighbourhood, loss of existing social ties due to neighbours' displacement, and changes at the municipal scale such as overall housing price increases that generate concern for families. Here, one may link 'displacement pressure' to what Davidson and Lees (2010) refer to as 'phenomenological displacement'. In this respect, those households under displacement pressure can be said to have been put in a state of on-going displacement.

The expansion of our understanding of how displacement can be multifaceted and nuanced allows us to see gentrification and the social consequences of neighbourhood changes in a profoundly different way. For example, official figures from governments or real estate developers involved in a redevelopment project mostly report the scale of last-remaining displacement by conducting head-counts of those tenants and owner-occupiers who are known to be occupying dwellings at the time of carrying out a neighbourhood census. Such an exercise involves boundary-making in terms of determining the rightful holders of legal property rights and identifying who qualifies for any kind of compensation measures for owners and tenants (see Ascensao 2015 on this in Lisbon, Portugal). For instance, in mainland China, such head-counting is usually reported by local governments and the media in terms of how many households are subject to relocation, and only includes permanently registered households confirmed by the official records kept by local neighbourhood committees. The immediate shortfall of such an approach is how the actual scale of displacement is hugely underestimated, excluding in particular private migrant tenants.

The underestimation of the scale of displacement also results from the long lead time of urban redevelopment projects, which place displacement pressures on local residents, prompting their house-moves even before actual redevelopment itself is carried out. This suggests that chain displacement is closely linked to development pressure in redevelopment projects that result in new-build gentrification. This is clearly demonstrated in a case study of Nangok's neighbourhood redevelopment in Seoul, South Korea, taken from Shin (2006, 2009a).

Case study: Nangok neighbourhood redevelopment

Nangok, located in the southwestern periphery of Seoul in South Korea, developed along a hillside, initially designated as a relocation site in 1968 for those households displaced from a number of sites in central Seoul. The majority (93 per cent) of the land on which Nangok neighbourhood sat was publicly owned, resulting in the absence of formal land tenure for most dwelling owners. During the 1970s and 1980s when South Korea was experiencing rapid urbanization and industrialization, Nangok saw an expansion of its population size, largely due to incoming migrants from rural regions and other urban poor families who found rents and housing prices in Nangok much more affordable for their household economy. By 1991, according to an official record kept at the municipal government, 2,732 illegal dwellings were present in the neighbourhood, accommodating 4,416 households or 16,734 people in total (Seoul Municipal Government 1991: 186).

When the municipal government was embarking on a city-wide redevelopment scheme known as the 'joint redevelopment programme' in the early 1980s (also see earlier in this chapter), local residents in Nangok also heard rumours that the neighbourhood was to be subject to such redevelopment. The rumours became a reality in May 1995

Figure 7.2: View of Nangok neighbourhood before its complete demolition, 2001 (Photograph by Hyun Bang Shin)

when the neighbourhood received conditional designation as a redevelopment district by the municipal government. The designation also signalled restrictions on further growth of the neighbourhood, including the prohibition of any additional investment in upgrading dwellings or building extensions. The prospect of imminent demolition and reconstruction of the neighbourhood also discouraged long-term security of tenure especially for tenant households. Nevertheless, the neighbourhood still retained a substantial number of residents whose size totalld 14,640 by 1996 (Gwan'ak District Assembly 1996).

The redevelopment of Nangok faced a number of setbacks in the following years particularly due to the Asian financial crisis in the late 1990s, which negatively affected the real estate markets in South

Figure 7.3: View of redeveloped Nangok neighbourhood filled with mostly commercial flats, 2007 (Photograph by Hyun Bang Shin)

Korea, discouraging the profitability of the neighbourhood redevelopment plan that depended heavily on raising profits from selling commercial flats to subsidize the construction of flats earmarked for original property owners. The original developer who was given the right to redevelop in 1996 had to withdraw from the project due to the bankruptcy of its holding company. Eventually, a public corporation intervened to salvage the project, and the displacement of the last-remaining residents started in October 2000. By this time, there were around 2,450 households (about 10,000 people). However, this meant that between 1996 and 2000, a little less than 5,000 people chose to leave the neighbourhood, most probably in part to avoid the inconveniences of demolition and in part as a result of property-owners' efforts to vacate dwellings. Official records that refer to the number of people affected by the neighbourhood redevelopment only refer to the last-remaining residents, but such records do not tell us the full picture and hide the occurrence of chain displacement that took place before the displacement of last-remaining residents. Nor do they show us the size of exclusionary displacement, that is, the exclusion of poor households who could have easily chosen those dwellings in Nangok as their most affordable housing options had the neighbourhood not been bulldozed.

MEGA-DISPLACEMENT: A GLOBAL SOUTH PHENOMENON?

Mega-displacement is often associated with cities in the global South, but it is also becoming a dominant trend in cities in Western Europe and North America, facilitated by the intense financialization of the real estate sector and the aggressive action of neoliberalizing punitive states to implement accumulation by dispossession (Harvey 2005). In particular, mega-gentrification has been associated with the process of redeveloping large-scale public housing estates in some Western European and North American countries (Bridge, Butler and Lees 2011). Public housing estates were formerly seen to be part of the decommodified means to resist gentrification (see Marcuse 1985b), but they are increasingly subject to territorial stigmatization (Wacquant et al. 2014) and the onslaught of gentrifying forces in the name of private-public sector partnerships and urban renaissance using mixed communities policy (see Chapter 5 this book).

For instance, the HOPE VI programme in the US, since the early 1990s, has initiated the redevelopment of deprived large-scale public housing estates in the US, which have become subject to redevelopment (read gentrification) (Zhang and Weismann 2006), and authors such as Lees, Slater and Wyly (2008) and Goetz (2010) have identified the deconcentration (read displacement) of the original residents as an outcome of the policy. Kingsley et al. (2003) examined 103 HOPE VI project sites that received grants up until 1998, and point out that '25,628 households were relocated from these sites from the time the programme began through the end of March 2000' (p. 429). This suggests that on average, 259 households had been subject to relocation from each site. Public housing estates subject to the HOPE VI programme are experiencing stigmatization, as they are characterized as experiencing 'severe distress' in terms of high incidents of family poverty, crime, vacancy and/or physical dilapidation (see Goetz 2010: 139 on eligibility for the programme).

The stigmatization of public housing has also been enacted in the UK where council estates have been placed under threat by a coalition of local councils and developers under the banner of creating mixed tenure through redevelopment (Lees, Slater and Wyly 2008). In London, in particular, the redevelopment of council estates in a similar manner to the HOPE VI programme has produced significant changes in both project sites and neighbouring areas (Lees 2014c). It has, in the words of Goetz (2010: 153), 'activated nascent land markets or swept away the last remaining obstacles to gentrification' (see also Wyly and Hammel 1999). As observed in Paul Watt's (2009) study of housing stock transfer policy in London, schemes such as the New Deal for Communities regeneration scheme that was to enframe the idea of community or public participation often turned out to be 'a Trojan horse for state-led gentrification in London' (ibid. 240). In the case of the Ocean Estate in Tower Hamlets, for example, the largest council estate in the borough, the local council promoted the transfer of public housing stocks to a housing association to inject capital for much-anticipated regeneration of the estate. This involved the demolition of 543 homes, comprised of 440 rental and 103 leaseholder properties. The demolition was followed by the construction of 714 private units that were meant to provide finance for the refurbishment of the rest of the estate's properties. No units were to be provided for the displaced tenants, while the hundreds of incoming wealthier families created displacement pressures and phenomenological displacement through the transformation of the estate. Lees (2014b, 2014c) discusses the significant and disturbing displacement of thousands of residents (council rent and leaseholder) from the Aylesbury and Heygate estates in Southwark to make way for newly redeveloped housing estates (see also Municipal Dreams 2014).

There are at least fifty council estates in inner London being affected by mega-gentrification and displacement; we estimate the direct displacement of tens of thousands of council tenants (and many thousands more if we were to factor in indirect or exclusionary displacement), in what is the removal of the final bricks in the wall of total gentrification for inner London (see Figure 7.4).

CONCLUSIONS

The Parisienne's experience of Haussmannization in the nineteenth century has become a lived experience for many urbanites in the contemporary global South, and it has also re-emerged for a faction of marginalized urban inhabitants in certain cities in the global North. The rise of mega-scale gentrification testifies to the consolidation of a new gentrification economy (see Chapter 3) and builds on the increasing proliferation of slum gentrification (see Chapter 6). Mega-gentrification also denotes a powerful nexus between the state sector and businesses, which feeds upon the developmental potential (i.e. rent gap) produced in a variety of urban settings (see Chapter 2). We have seen earlier that such rent gap formation does not always depend on disinvestment in gentrifiable properties, but also on growing affluence in adjacent urban districts that outpace, for example, the growth of the developmental potential in substandard settlements.

Figure 7.4: Displacement, Heygate Estate, London, 2012 (Photograph by Loretta Lees)

The experiences of mega-displacement and gentrification testify to the importance of looking at the complex sets of interests that are shaped and reconstituted in conjunction with the changing geography of property-based interests. The hegemonic power of the state and businesses does not enable mega-gentrification simply through the use of violence and coercion. In Seoul, the proliferation of substandard settlements was initially subject to forced eviction and demolition by the state throughout the 1970s (see Mobrand 2008 for an historical account of this process). However, their eradication was only possible when the state devised a redevelopment programme (JRP) that enticed indigenous property owners to join property-based interests. Embedded in a speculative real estate market, these interests aimed to turn these neighbourhoods into a higher and better use, turning the land on which the substandard settlements sat into profitable commodities. The history shows that poorer factions of these property owners also lost out in the process (see Shin 2009a; Ha 2001); indeed, the separation of property owners from 'property-less' tenants was a classic divide-and-rule strategy to weaken opposition to the state-led redevelopment programme. A similar situation is also apparent in the process of speculative urban development in mainland China where private tenants, who are the most severely affected by mega-gentrification, get excluded from the outset. Property owners' claims on their properties in urban neighbourhoods secure legitimacy only to the extent that they are recognized by the local state that acts as a *de facto* landlord (Shin 2009b).

The cases of mega-gentrification discussed in this chapter show us that the experience of displacement differs across different segments of the population in gentrifying neighbourhoods. This means that it would be dangerous to identify displacees as a single, homogenous group of residents with a uniform interest. As Sapana Doshi (2013: 845) states in her discussion of slum redevelopment in Mumbai, 'uneven displacement practices are central to the social production of land markets'. As such, 'slum redevelopment [in Mumbai] has privileged local and transnational elites through a combination of negotiated consent to displacement and forced eviction' (Doshi 2013: 848). The use of consent and coercion is at the core of Gramscian understandings of ruling class domination (Gramsci 1971: 57–8), and as Haugaard (2006) summarizes, the hegemony of dominant interests is secured through 'a series of consensual alliances with other classes and groups' (p. 5). For mega-gentrification and displacement to continue as a dominant mode of urban accumulation, it seems necessary that the manufacturing of a broader societal consensus to maximize opportunities of profiteering from real estate investment be built on a set of coercive measures to dissipate discontented voices. Such consensus building is sometimes facilitated by the dystopian imagery of criminalizing squatters (e.g. in Taipei) or welfare dependants (e.g. in the UK and the US) by increasingly neoliberal[izing] states (see Jou et al. 2014; Lees 2014b; Wacquant et al. 2014). Sometimes, as East Asia's history of speculative urbanization under developmental states attests to, the proliferation of speculative property-based interests in the state sector, populist sector, and the business sector, seems to have created a particularistic culture and discourse (cf. Ley and Teo 2014 regarding Hong Kong and Shin and Kim 2015 regarding South Korea) that have increasingly excluded the urban poor.

8 Conclusion

Ruth Glass's maiden name was Lazarus, and perhaps inevitably, the term she coined has refused to die out; indeed, it has been reborn as planetary gentrification. Schulman (2012: 14, emphasis in original), like we have done in this book, reflects:

> What is this process? What is this thing that *homogenizes* complexity, difference, dynamic dialogic action for change and *replaces* it with sameness? With a kind of institutionalization of culture? With a lack of demand on the powers that be? With containment?

Her answer to this question, like ours, always comes back to the same concept: *gentrification*. As she goes on to say:

> First I needed to define my terms. To me, the literal experience of gentrification is a concrete replacement process. Physically it is an urban phenomena: the removal of communities of diverse classes, ethnicities, races, sexualities, languages, and points of view from ... cities, and their replacement by more homogenized groups. With this comes the destruction of culture and relationship, and this destruction has profound consequences for the future lives of cities. (p. 14)

We agree, and our book makes an important contribution to collective thought by building on recent postcolonial critiques of urban studies from different parts of the world to develop a more planetary view of gentrification. It has been recognized for some time now that English language (and indeed non-English language) gentrification debates have been narrowly restricted mainly to studies of North American and West European cities. Until recently, there have been only a few sporadic attempts to address this bias, but times are changing and there is now a significant and growing gentrification literature (English and non-English speaking), focusing on cities outside of the global North in which researchers are trying hard to think about non-Western gentrification differently. This is because academics internationally are realizing that gentrification helps us understand what is happening in the spatial restructuring of different cities, under different (capitalist) regimes. In producing this book, we have made every effort to draw on newly emerging gentrification literatures and ideas, but we have also relied on, as stated in the introduction, the high-level discussions held at two international workshops on gentrification (in London and Santiago), and the detailed discussions that occurred in the editing of another book (Lees, Shin and López-Morales 2015); additionally, we have worked as an editorial team on two special issues on global gentrifications in East Asia and Latin America. Based on this work, this book offers a new look at the critical political economy of planetary gentrifications.

In reflecting on reversing the flow of intellectual authority in comparative urbanism, Gareth Myers (2014: 116) suggests that: i) it is not simply about starting in the global South or turning-around the telescope to look from non-usual contexts, or ignoring the usual suspects and doing comparative urbanism without the West, 'the key lies more in placing cities on a level analytical plain in comparative studies'; ii) the mobility and circulation of urban policy remains an important focus; and iii) more comprehensive and multiregional comparison is only possible through multicultural research teams and networks of researchers. This book constitutes a concerted attempt to place gentrifications globally 'on a level analytical plain': urban policies and programmes of renewal or redevelopment have been a significant focus, and we have been a multicultural and globally connected research team. In *Planetary Gentrification*, we have been forced to ask ourselves many questions: Are we expanding the term gentrification so it means nothing? Or, has gentrification coped with the test of comparative urbanism? What is essential about gentrification as a concept? What does it do analytically that nothing else can do better? Is gentrification a useful concept outside the West for academics and/or activists? Who are the agents of gentrification now? Can you have displacement of the urban poor that is not gentrification? And so on. Indeed, we have asked ourselves throughout this book, what this means for Neil Smith's (1996) argument, nearly twenty years old now, that gentrification is generalizable:

> [G]eneral differences really do not get into a sustainable thesis that these [instances of gentrification] are radically different experiences . . . the existence of difference is a different matter from the denial of plausible generalization. I do not think that it makes sense to dissolve all these experiences into radically different empirical phenomena. (pp. 185–6)

We have considered whether the concept of 'gentrification' has global application, and whether there really is such a thing as a gentrification generalized. After much research and international discussion, after 'distinguishing between conditions of possibility or enabling conditions applicable across the board and contingent factors addressing specific contexts' (Betancur 2014: 1), after discussing, dialoguing and comparing the cases we have brought to this book, we have concluded YES – as long as we keep gentrification general enough to facilitate universality while providing the flexibility to accommodate changing conditions and local circumstances. Indeed, to keep pace with the new geographies of gentrification, we need a gentrification theory that is both located but also dis-located. Betancur (2014: 2) identifies a structural core of gentrification or universal conditions of possibility (enabling conditions), which include: societal regime shift and the associated restructuring of cities, encompassing key factors such as 'rent production, reproduction and capture, the production of gentrifiers and gentrifiable areas; and class displacement/replacement'. And *contingent/conjunctural (necessary) factors* that determine gentrification's 'actual presence, form and possibilities', 'accounting for variations by geographies' and temporalities (ibid.). He suggests that if we were 'to study gentrification in nontraditional contexts, then you should start 'with regime shift and corresponding enabling conditions to then identify contingencies' in different localitie (ibid: 3). Betancur (2014: 9) argues that both the enabling and necessary factors 'have to concur at a critical scale and be brought together by social actors turning potential into actually existing gentrification'. Those actors or agents, we argue in this book, are now, globally, less likely to be Ruth Glass's (1964a) gentry or David Ley (1996) and Tim Butler's (1997) left liberal pioneer gentrifiers. They are more likely to be the state (twinned with corporate capital) and powerful local or international developers. In the case of China, state powers are intertwined with an emerging, increasingly affluent (also tremendously powerful) new middle class. In the case of Rio de Janeiro, Brazil, the whole redevelopment process in the city is underpinned by transport-oriented public investment penetrating the most emblematic favelas in order to increase their 'spatial capital', as we saw in Chapter 3; also transnational elites operate there to grab land and properties in order to undertake profitable tourist-oriented projects. There are peculiarities to state-led gentrification in the global South and there are different experiences of gentrification around the globe, as illustrated in this book, but these are not radical enough to warrant dilution or dismissal of the term 'gentrification' (also see Smith 1996). Indeed, when we stretch the term gentrification at a planetary scale, we are operating with an abstraction or conceptualization of gentrification, focusing on the necessary relations of gentrification centred on the commodification of space and unequal power relations, often pre-existing, in any society (see Chapters 2 and 3) (also see Clark 2005). As such we are able to develop 'knowledge, understanding and generalization at a level between what is true of all cities and what is true of one city at a given point in time' (Nijman 2007: 1). How gentrification occurs in a concrete manner in a given locality requires researchers' investigations to unravel contingent relations (e.g. state regulations, the rise of middle-class groups, civil society formation, subaltern urbanism, informality) to highlight the particular path of 'endogenous gentrification'. In doing so, we believe that the rhetorical power of gentrification increases rather than subsides. Our approach also indicates that gentrification may exist as an embedded process in places where it may experience epistemological absence, as Ley and Teo (2014) have shown with regards to Hong Kong. After all, epistemology is a matter dependent on who researches, for the sake of whom, and based on ontological principles of social justice.

Gentrification, as a process, remains (despite global recession, and in some places like in the US today 'due' to global recession) the eye in the urban capitalist storm pushing capital into all sorts of planetary urbanizing environments which were not fashionable or in some cases even conceivable before. We have agreed with the thesis that suburbanization is growing around the world on megacity peripheries but we also have argued here, and extensively corroborated, that urbanization around the world is seeing the production of multiple centralities and that processes of redevelopment are at the forefront of conflictive political arenas everywhere, as land and transport costs increase around the globe due to financial speculation increasing global demand. We do not necessarily claim that gentrification is the most dominant urbanization process in the world, but we do argue that without gentrification studies, the processes of urbanization in the world cannot be understood in all their complexity; indeed, the problems arising from class-led redevelopment are among the most salient urban problems on the planet.

In this book, we ask our readers to reflect critically on these gentrifications which are causing increased social inequality globally. *We ask them to think about planetary urban futures under conditions of gentrification.* We demand that gentrification studies steps outside of the 'West', and in so doing the chapters in this book provincialize and contextualize the hegemony of Anglo-American case studies. But as Lees (2012, 2014a) argues, we also want gentrification studies to return to the West again to reconceptualize hegemonic theories and concepts using the insights learned from outside of the West. Our comparative urban approach seeks to theorize back, reflecting, as we do throughout the book, but especially in this conclusion, on what this exercise in a comparative urbanism of gentrification means for existing theories on gentrification and for the notion of a generalized, global gentrification.

The comparative urbanism of planetary gentrification that we have undertaken in this book has drawn inspiration from Robinson (2003, 2006, 2011c) and Roy (2009), and in tandem with Lees, Shin and López-Morales (2015), has sought to prize open gentrification studies' Eurocentrism. In so doing, we hope to give both new vigour and new analytic rigour to gentrification studies. Although we have been, and remain excited by, the emancipatory epistemological potential of exploring cities outside of Anglo-America, especially those in the global South, on their own terms (e.g. McFarlane 2010; Bunnell et al. 2012), our critical urban geography is a little different from other comparative urbanists. We believe that to flatten the globe and its multiple urban hierarchies with an appreciation of difference hides social injustices and neglects power relations (Lees, in press), which are very apparent in the process of capitalist accumulation that entails commodification of space in particular. For us, gentrification in its 'order and simplicity' (Clark 2005) remains at the focal point of this process.

Under planetary urbanization that extends the capitalist forms of urban restructuring to every corner of the globe, gentrification inevitably becomes part of urban place-making. For us, Neil Smith's proposition of gentrification generalizable is not a proposition that imposes Western notions of urban theorization on the global South, but the realization that this fundamental capitalist logic of accumulation and inequality, has not become but is, global, affecting the daily lives of those whose life chances are increasingly shaped by the ways in which the built environment is exploited by the state and capital. Like with Spivak's (2003: 71–102) 'planetarity', we too seek to overwrite the globe and globalization with respect to gentrification studies. In so doing, we seek to displace and realign the axis of 'comparison' for gentrification from *Euro-America* or the West, not to the global South, but to *the planet*. The book represents a conceptual-intellectual effort to look at gentrification in and through quite different contexts, and to pull out explicit and implicit comparisons (also see Jazeel and McFarlane 2010: 118; Robinson 2011c: 129). In so doing, we critically engage with both academic globalization *and* the globalization of capital. But at the same time, we are interested in somewhat different (if interrelated) sorts of injustice. For most of the new comparative urbanists, it is the injustice of neglect and misrecognition of certain (Southern/Third World) cities: for us it is the injustice of class exploitation. It is also the injustice of neglecting or misrepresenting the insights of local researchers and their contributions in explaining their own urban realities in processes of class-polarized urban change. To directly address this issue, throughout the book we have drawn extensively on the research and thoughts of others – whom we would like to both acknowledge and thank in this conclusion. It is our contention, however, that

> these two injustices cannot be separated and that we need now to find a theoretical/conceptual and methodological way forward that has

political punch. We do not simply want to transcend the oppression of location but of human beings being cleared out of cities world-wide, socially cleansed simply because they do not have the money, the power, or the face that fits the new urban world. (Lees, in press)

WHAT WE HAVE LEARNT

Not surprisingly we have learnt that planetary gentrifications are complex. The peripheralization of low-income residents (Wu 2004; Fang and Zhang 2003) happens in some places, but not in others; indeed, with the presence of multiple centralities, peripheralization itself needs to be redefined. This however is beyond the scope of this volume. Gentrification is now increasingly associated with the formalization of housing and labour markets (Winkler 2009). This has led to slum gentrification (see Chapter 6) and the revanchist retaking of central city spaces, especially in South America. Much of this formalization is associated with repression (Gaffney forthcoming) and a gender and ethnic revanchism in city centres (see Wright 2014; Swanson 2007). Academics and activists in the global North may learn more from the revanchisms in the global South than from those, such as New York City (e.g. Smith 1996), that have become hegemonic in the gentrification literature. The revanchisms are all a little different – in Mexico, it is about the state trying to control civil society, which it views as out of control; in Pakistan it is the army who are revanchist. There are questions about what revanchism *is* in the context of non-democratic societies, e.g. those under military control. Rio de Janeiro's *Policiais Pacificadoras*, the Special Group for Urban Control, who have suppressed slum dwellers and street vendors, are trained by the army and use military weapons.

Our concern in engaging with a wide range of cities was that this would create another kind of 'imperialist appropriation' (Robinson

2011c: 19) of urban experiences in the service of Western scholarship – and bringing back ideas to the global North has the sense of this. But it is rare that academic literatures from the South get to present lessons to the North. Here we give gentrification studies that opportunity.

Gentrification is led now by an alliance of national and local states and capitalist developers, as we have seen in Santiago, where redlining and ground rent dispossession has been enacted to create markets for international developers (López-Morales 2010, 2011, 2013a,b). In Asian states, like China and South Korea, wholesale metropolitan restructuring based on comprehensive housing relocation programmes have caused mega-displacements (Fang and Zhang 2003; Shin 2009a,b; Shin and Kim 2015). In Rio de Janeiro, the whole urban system, since the early 1990s, has surrendered to a state-led process of service-oriented economic reshaping; this has been aimed at reversing the decline of its post-industrial urban economy, but it is causing significant displacement of *moradores* from the favelas they have built and occupied for decades, and in some cases for a century (Queiroz Ribeiro, 2013). Much of this has involved the restructuring of tenure to promote home ownership both in the global South and North. There has also been significant public investment in the amplification of spatial capital (Rerat and Lees 2011; Blanco et al. 2014), and it is here that gentrification studies could return to early work on centrality and update those economistic theses (as we argued in Chapter 3).

As we recall Neil Smith and agree with his thesis that acknowledging difference is not the same as denying plausible generalization, we have discussed here an ample array of emerging cases and theorizations to see what is new about gentrification in the world. The politico-economic aspects of the phenomenon are at the forefront of our discussion, as new state-business coalitions have taken control of extensive parcels of land and new, very powerful spheres of governance allow capital to flow into and switch from local spaces to global spaces. This has been achieved using an array of policy tools, such as zero tolerance policing (see Chapter 5); our argument is that this is happening today at a much larger scale and with more aggressive state-policy orientations than 30 or 40 years ago.

The new evidence we have analysed for this book sheds light on the regularities of the capitalist system to constantly accumulate the riches accrued from urban land and monopolize what comes to be the social production of economic space, something that Smith once called 'potential ground rent'. At the time, that concept was a mere 'predictor' of gentrification that might happen, but we know today the political complexity of the 'potential ground rent' concept offers more than that. Indeed, after decades of debate in the North about whether the rent gap thesis was useful enough to predict gentrification in what hitherto constituted an exclusively 'Northern' scenario of agent-led gentrification, it seems today that the rent gap is not only relevant to measure diverse forms of land exploitation, stigmatization, displacement and exclusion, and the outcomes of many urban policies for redevelopment, but also to know to what extent capital inflows and outflows transform social spaces; insofar as the concept of creative destruction (which plays out in gentrification) implies an everyday struggle between intensive capital-led destruction of local environments and the social forces occupying those spaces under dispute. And, the importance of capitalist crises in the constant cycles of devaluation and revaluation of landed capital can be underlined again, this decade, reminding us how important it is to understand urban processes from a systemic perspective. In this vein, the emergent global gentrification literature offers an interesting analysis of current urban capitalism, it tells us how the system is currently working and mutating, and this can be seen relationally from the bottom and the top of the social ladder and power structures. This is what gentrification studies has to offer: relational perspectives that can hardly be found in other conceptual spheres.

FROM SOUTH AND EAST TO NORTH AND WEST

> What new, indigenous or cosmopolitan theorizations can be brought to bear on gentrification in the global South, and in turn the global North? (Lees, 2012: 166).

In her 2009 paper on 'new geographies of theory' in our contemporary world, Ananya Roy called for 'a more contoured knowledge of its cities' so that imaginations and understandings emanating from the analysis of urban processes in the global South can enrich our understanding of urbanizing processes 'in *all* cities'. In similar vein, Gareth Myers (2014) also attempts to see how American cities can learn from African cities. It is well worth contemplating what cities in the global North/West can learn from cities in the global South/East and from researchers based in those places with respect to gentrification.

As we have discussed, outside the global North, gentrification is happening at the same time as suburbanization or peripheralization, rather than as a reaction to it. Rather than a 'back to the city' (from the suburbs) movement by capital and people, a notion that governed gentrification researchers' enquiries in the West for many decades, gentrification outside the global North is a process that cannot be dissociated from urbanization itself, which accompanies suburban expansion as well as intensification of the use of existing urban spaces. Planetary gentrification is not a reaction against suburbanization: indeed, the two processes are increasingly blurred, as they are in the global North now too (also see Lees 2003 on the suburban mindsets of super-gentrifiers and Butler 2007b on suburban gentrification in London's Docklands). Especially in the context of the speculative urbanization that is rampant in many cities of the global South, the state itself has accommodated the demands of gentrifiers, and at the same time, been implicated in the production of these same gentrifiers and their desires (also see Ley and Teo 2014). The urban as a way of life has been very much emphasized by a number of states in the global South, mobilized as an ideology that justifies intense investment to facilitate urbanization, a process that makes investment in the built environment as important, if not more so, than industrial production. The ideal of urbanization and not 'back to the city' may well have more purchase now in the global North too, where the ascendancy of speculative real estate capital has been witnessed with the help of the financialization of real estate commodities, despite the inherent contradictions that produce periodic financial crises.

Like in the West, the global South has witnessed the rise of emerging 'new' middle-class populations as 'gentrifiers'. In some countries their 'new' middle classes emerged at the same time as those in the West; in others, they are a more recent addition to society. In the global South, their emergence has been related to different temporal paths of modernization and growth, whether under post-socialist or developmentalist states or market-oriented regimes (see Chapter 4). 'New' middle-class gentrifiers around the planet cannot be viewed as a singular 'global gentrifier' class (Rofe 2003), for they are quite different in different places. In some locations, they are relatively very well off; in others, gentrifiers can be marginal-income groups. In other countries, the 'new' middle classes are a fragmented group, some sharing the fortunes of economic success in their respective economies, accumulating real estate assets through homeownership, while others are less well off. Irrespective, speculative urban accumulation has provided a positive view of redevelopment (read gentrification), 'shaped by a widespread popular belief in the role of property in upward social mobility' (Ley and Teo 2014: 1300), even if this belief is seeing some signs of fatigue more recently.

The complex identities of contemporary gentrifiers are very apparent in mainland China. Its land and housing commodification has meant that urban citizens have a 'choice' now about where they want to live, dependent of course on their wealth, whereas before people were allocated homes by the state. In addition, the One Child Policy in China and the continuance for now of traditional family structures must be brought into the equation, especially in terms of how inheritance enables the younger generation to access homeownership. The intervention of the modernizing state has created the conditions for gentrification and as such led to demand from gentrifiers: indeed the state has produced gentrifiers, while the state itself has become the leading gentrifying force. The emerging 'new' middle class in China is not a singular entity: it is a much differentiated class whose lifestyles range along the continuum from conservative and traditional to liberal. Ren (2008) argues that in Shanghai, the middle class is less motivated by proximity to work or by a frontier spirit, and more by expectation of financial return from the appreciation of their properties, but this may be less the case in other Chinese cities, outside of Shanghai and Beijing. The classic Western gentrifier does not map neatly onto gentrifiers in the global South. As Rose (1984) pointed out some time ago, the characteristics of gentrifiers need to be constantly reviewed and reinvestigated as gentrification changes and in different contexts (see Lees, Slater and Wyly 2010: 218–19).

In the book, we have underlined the importance of analysing displacement with respect to planetary gentrification. In conclusion, we want to underline the importance of previous histories of displacement in different cultures, and consideration of how they come to bear on the thoughts and feelings about contemporary displacements. For example, in China, the Cultural Revolution in the 1960s and 1970s caused big displacements that are still quite fresh in people's minds. How does this impact how the Chinese think about gentrification-induced displacement? There are still questions about the legacies and lasting influences of Mao to post-Mao urban development for gentrification researchers working in/on China. Similar questions can be raised for other emerging and developing economies where more authoritarian states have produced mega-displacement through slum clearance and infrastructure projects, rendering displacement as something in people's daily experience. In the global North too, gentrification-induced displacement led by the state is now more significant in volume than the smaller displacements described by Ruth Glass. The importance of thinking back to earlier displacements is written into Derek Hyra's (2008) labelling of the state-led gentrification of public housing projects in the US as 'new urban renewal' – taking us back to the stories about displacement by the federal bulldozer in US inner cities in the 1950s and 1960s.

In the book, we highlight the 'slum gentrification' that is happening globally (Chapter 6). This stands against the assertions of some development geographers who have argued to our faces that there is no such thing as slum gentrification, especially in Latin American cities. Our book shows otherwise. But more importantly, what lessons can be learned from slum gentrification in the global South for slum gentrification in the global North or vice versa? One lesson is that whether the slum is formal (e.g. council estates in London or state-built housing areas in South Africa or Chile) or informal (e.g. favelas in Rio de Janeiro or slums in India), they can both be subject to gentrification if the context is right. In the process, these areas are designated as slums, even if their residents do not feel they are living in a slum (e.g. those on council estates in London). The term 'slum' needs renewed analytical attention, and 'slum geographies' from/in the global North and South need to talk to each other. We need better theorization of our new world of slums. Future research needs to pay attention to the role of informality, the fluidity between formal/informal processes and land uses, the issues around title deeds versus not, and so on. The increasing trend for the gentrification of slums in the global South/East is mostly due to state goals of modernization and the needs of capital to invest in new economic areas, and this constitutes a threat to socio-spatial stability in particular cities and in whole urban systems. The role of the state, alongside international capital and NGOs, cannot be underestimated, as part of the politico-economic repertoire of slum gentrification today. Gentrification researchers need to develop a clearer understanding of the circulation of capital in gentrification and realize that the agents that most benefit from state investments in slum infrastructure are the local stakeholders operating within and across slum areas, and that this destroys slum dwellers' lives (Desai and Loftus 2013).

In a strange turnaround, while some parts of slums (informal settlements) in Asia, Latin America and Africa are getting gentrified, informal housing in the form of garden sheds or shacks is on the rise again in cities like London as a result of gentrification and escalating real estate values (see Figure 8.1). In London, poor immigrant families have been found living in small garden sheds, rented out by unscrupulous landlords exploiting the housing crisis:

> Hidden in a cluttered courtyard behind an off-licence in Forest Gate, East London, is a narrow brick building no bigger than a shed that a family of four calls home. Asah (not her real name), her husband and two young daughters all share this ramshackle structure, built illegally by its owner to make money from desperate tenants such as her. Thousands of similar buildings exist across the country – many occupied by immigrants with little choice but to live in these dangerous and cramped conditions – and more are built every day. They are described by housing campaigners as Britain's 'modern-day slums' or, more euphemistically, 'sheds with beds' . . . Asah, 37, found her accommodation in a shop window after a long, fruitless search. She pays £375 per month for the small, two room shack. Rubbish and large breeze blocks are strewn across the courtyard outside her door, where her young daughters play. 'I know it is not good for us to be here. But it is enough

for us and we can afford it,' she says. 'My kids go to school nearby. We have tried looking for other places, there is nothing.

The irony of slum gentrification in the global South in tandem with the emergence of new slums in the global North does not escape us. Although Asah's story happened in London, it could have been recorded in any slum settlement in any city of the South. In some ways, this demonstrates that planetary urbanization today is far more a story about the complexities of the urban production of inequality and social polarization than just a story of growth, urban development and 'good practices'. Or maybe it is the overarching and unprecedentedly dominant current capitalist power of urban transformation

Figure 8.1: London's modern day slums (Courtesy of *The Independent*, 2015)

that is bringing cities in the South and North into essentially similar paths of redevelopment. In fact, we see here 'gentrification' or whatever the term that is used to label the displacement of low-income populations from recentralized spaces, 'serving' as a process that enlarges inequality and polarizes societies even more than they were historically.

But perhaps this temporal coming together of related processes planetarily is not as new as we might think. For state-led or privately led corporate interests in large-scale redevelopment existed in cities of the global South (like Seoul) earlier, or at the same time as, in the global North. For example, Davidson and Lees' (2005, 2010) 'new build gentrification' was happening in Seoul in the 1980s, before it took off in the UK and the US. We might even be able to trace a reverse story of large-scale, new build gentrification starting in cities in the South and then moving towards spaces in the North, but we suspect the story is more simultaneous than that.

Even though, for the most part, various gentrification blueprints have been imported from the North through international agencies, consultants and policy mobilities, like zer-tolerance policing to Latin America, the ways and means of gentrification are rather fluid and differ between cities of the North and the South. Mixed communities policy was being used as a tool for social stabilization in Calcutta in the 1970s in a similar vein to that in the global North today. The success of gentrification in remaking new and emerging urban centralities is dependent on levels of development, the penetration of global forces, the maturity of local real estate markets and economies, and the extent to which local markets are linked into global circuits of investment. There does not have to be a neoliberal state for gentrification to occur, as a range of capitalist regimes may have an equal stake in exploiting the built environment and its inhabitants through the commodification of space, but the registers of neoliberalism at play would certainly intensify the likelihood of gentrification.

RESISTANCE TO PLANETARY GENTRIFICATION

Resistance to gentrification is not the central focus of this book, despite some discussion of it in Chapters 5 and 6; nevertheless, gentrification has been (and is being) contested by numerous grass roots groups around the world. As such, in our conclusion, we want to (a) underline the importance not only of analysing gentrification but of resisting it too; (b) we would urge here, as an extension of our project, a comparative urbanism of anti-gentrification resistance. Part of this battle is about coming up with realistic alternatives for around the globe, which we turn to in the next section.

As this book has shown, gentrification is the appropriation of land to serve the interests of the wealthy. The question remains: who are cities for and who should we develop cities for? These are questions of political accountability. Despite gentrification being a significant planetary process, very little has been written about resistance to gentrification and even less about successful resistance. But in researching planetary gentrification, it became apparent that resistance to gentrification in the global South is: (a) not new, and (b) sometimes more successful than in the global North. Resistance has mostly occurred at the neighbourhood or city scale, but some have become national, e.g. the Abahlali movement in South Africa (see Cabannes et al. 2010; Lees 2014b). This is interesting in itself with regards to earlier statements by Hackworth and Smith (2001), who argued that resistance to gentrification was all but dead, but even more so when we consider that resistance to gentrification was happening in South Korea before gentrification authors in the West even began to discuss a global gentrification! Resistance to gentrification emerged in South Korea in the 1980s; it began to organize systematically and was supported by other social movements such as the democracy movement and the labour movement (see Kim 1998; Ha 2015; Shin and Kim 2015). In South Korea post-2008 this resistance has both increased and mutated in terms of the main actors, the form of protest, and interaction with other social movements.

Resistance to gentrification made international headlines in Taksim Square in Istanbul, Turkey, in 2013 (see Angell, Hammond, and van Dobben Schoon 2014). Not since the Tompkins Square Park riots in New York City (see Smith 1996) had anti-gentrification resistance made headline news. The protests began when a small group of activists tried to stop the demolition of Gezi Park – one of Istanbul's few remaining green spaces in the larger Taksim Square (see Figure 8.2); but the protesters were also unhappy with the rapid pace of urbanization in Turkey's metropolitan cities. They criticized the neoliberal turn of Erdogan's government towards the gentrification process that Istanbul was undergoing, arguing that the plans hit the neighbourhoods that house the poor, immigrants, Kurds, and the Roma hardest. When the police responded with extreme violence, the protests grew, with millions of demonstrators on the streets and protests held in dozens of cities throughout Turkey. More than 8,000 people were injured, six demonstrators and one policeman were killed. The struggle for Gezi Park further intensified and diversified unscripted encounters, transforming a public park – through the collective work of anonymous urbanites – into a commons (cf. Harvey 2012: 73). Resistance to gentrification was not, however, new to Istanbul. İslam and Sakızlıoğlu (2015) discuss in depth examples of earlier resistances in the neighbourhoods of Sulukule and Tarlabasi.

In mainland China, urbanites are increasingly mobilizing around what could be termed as *weiquan* (defending rights) movements, especially among homeowners who are keen to protect their quality of life and property values in their new housing estates. Residents' protests against pollution-generating facilities have been on the increase, and so are struggles by rural villagers against the expropriation of village lands. In a neo-authoritarian political context, such collective mobilization of homeowners and villagers has received attention among critical

Figure 8.2: Anti-gentrification rally, Gezi Park, Taksim Square, Istanbul, 2013 (Photograph courtesy of Tolga Islam)

scholars for its role in contesting state hegemony and state-led urbanization (see Chen and Webster 2005; Hess 2010; Lee 2008; Li and O'Brien 2008; Yip and Jiang 2011). However, it remains to be seen if the rise of the new middle classes in mainland China provides a favourable situation in which struggles against forced displacement can be fought. What the Party State imagines is well-off, new middle classes who are increasingly detached from the working classes, villagers and the poor (Shin 2014a: 513–14). Given this, there is some purchase in the idea of re-theorizing the right to the city idea away from the global North. Unlike the classic American anti-gentrification battles over the 'right to stay put' (Hartman et al. 1982, Hartman 1984), in Chinese cities, the fight is over location, resettlement housing or monetary compensation. The rights claims tend to be framed both by the powerful and the weak, as struggles over economic subsistence and security, rather than against the state itself, although the intensification of economic struggles could evolve into political ones (Shin 2013). China's urban citizens are usually outweighed by the state that is the legitimate owner of urban land, while urbanites with homeownership are in possession of land use rights for a limited period of time (maximum seventy years for residential properties). Claims by the state on land and housing usually precede individual's rights claims. In addition, all of the disputes are affected by the unequal power relations that prop up institutional and legal frameworks; legal procedures for determining compensation and relocation are often contradictory or ambivalent, creating various loopholes that serve the interests of the state and businesses more favourably (see Shih 2010; Zhang 2013). Institutions and laws discipline such dissent rather than enabling rights (see Sakızlıoğlu and Uitermark 2014 on the institutionalization and internalization of displacement rules).

There are a number of in-depth studies of resistance to state-led gentrification in the global South that we would urge our readers to look at: for example, the resistance in the case of the Las Gladiolas public housing project in Puerto Rico (see Morales-Cruz 2012), that we discussed in Chapter 5; the case of barrio women who successfully fought off an 'urban renewal' project that was to gentrify the Old Market in Caracas (see Velásquez Atehortúa 2014), for there are few, if any such successes in the global North; and the case of the Pakistan Mahigeer Tahreek (the Movement of the Indigenous Coastal Fisherfolk Communities of Pakistan) who fought against a Dubai-based developer in Karachi, Pakistan, who subsequently backed out of a large real estate development (see Hasan 2012). Betancur (2014) argues that unlike in the North, gentrification in Latin America has also run into stubborn resistance from the self-help/self-employment spaces that the lower classes live in. As a result, gentrification has been much more limited than expected (see Ley and Dobson 2008, on gentrification limited). In Lima, residents organized around the Comite Promotor para la Renovacion Urbana with 'Renovacion urbana sin Desalojos' (Urban renewal without evictions) (see International Alliance of Inhabitants 2008) and in Colombia opponents of gentrification named the Office of Urban Development the Office of Urban Displacement!

We do not argue that social movements in every locality need to introduce 'gentrification' into their vocabulary or slogans, for it has been acknowledged that gentrification as a material process has been existent despite its epistemological absence (see for example, Ley and Teo 2014; Shin and Kim 2015). However, in situations in which displacement is regarded as an inevitable by-product in the pursuit of public interests, it is important to dismantle the myth that equates redevelopment with societal progress (also see Tang 2008), and for displacement, and therefore gentrification, to be recognized as a negative process that needs to be tackled by all means. Without official recognition it is harder to resist.

ALTERNATIVES TO PLANETARY GENTRIFICATION

Critical urban learning involves questioning existing urban knowledge and formulations, but it is also about proposing alternatives – not just conceptual and epistemological alternatives but also alternatives to gentrification. There has been increasing discussion of resistance to gentrification (e.g. Janoschka et al. 2014), but academics rarely, if ever, discuss alternatives to gentrification. Of course, Marcuse (1986) asserted the decommodification of housing as the alternative, but in a situation of planetary gentrification, is the decommodification of housing (in the Western sense) a viable alternative? We can think of a number of possible alternatives to gentrification, which range from local government protection and expansion of affordable (social) housing, community land trusts to guarantee affordability forever, the introduction of anti-speculation taxes to rent stabilization, limited-equity housing cooperatives for renters and urban policies to capture increased land value, which can in turn be channelled into social justice programmes. But first we need to understand that the phenomenon of speculative landed interests does not exist in the same way everywhere. We need to know much more than we currently do about how opportunities for profit from real estate are produced in place, as well as the timing, extent and agents behind speculative investment in real estate. We would argue that grassroots everywhere could learn from deeper gentrification research.

Devising localized action plans to challenge planetary gentrification and present alternatives requires critical scholars and activists to rethink how existing institutional systems can be used to their advantage. The nature of the state and its relationship with civil society is going to be an important factor that influences the ways in which alternatives to gentrification can be thought of. Back in 2011, one of the authors was attending an international workshop on urban utopianism held at Hong Kong Baptist University, which brought together a number of critical scholars from Asia and Europe. An interesting debate took place in the midst of the workshop, which touched upon the issue of how the state is to be understood. While most scholars particularly from Southeast and East Asia launched heavy criticisms against the state, an eminent scholar from Scandinavia raised the question, to provoke the audience, about the possibility of working with the state, rather than sidelining it in thinking of alternatives. Such a debate reflects the long-standing position of the state in each region, the Asian state, especially (neo-)authoritarian and developmental states in East Asia, have suppressed civic rights and undermined democratic principles while legitimizing themselves through economic growth and their ability to feed their nations. This is in contrast to the social democratic states in Scandinavia, where procedural democracy and participatory planning enable the active participation of civil society groups and grassroots organizations. Then again in Sweden some activists are wary of the Scandinavian, neoliberal, privatizing state; for example, a group of Swedish academics and activists have used The London Tenants Federation, Lees, Just Space and SNAG's (2014) booklet *Staying Put: An Anti-Gentrification Handbook for Council Estates in London* and rewritten it for different Swedish cities from Gothenberg to Stockholm. In this, they assume the state as a gentrifier, not a viable solution to gentrification. It may be that anti-gentrification handbooks need to be developed for specific cities and circumstances world-wide.

To that end, we would urge scholars and activists alike to identify the contingent factors that define the trajectory of gentrification in a given locality, determining the particularities of gentrifying forces in neighbourhoods, municipalities, regions and countries in order to develop place-specific anti-displacement and anti-gentrification agendas and action plans. No matter how ripe economic conditions are for gentrification to occur, it is the political struggles by agents that determine the final fate of urban inhabitants in gentrifiable urbanizing spaces. These struggles inevitably call for localized action plans (rather than 'one-size-fits-all' solutions), aimed at challenging the state hegemony that increasingly has a stake in capital (re-)investment in the built environment, whose mode of existence needs to be widely diversified for the survival of the state (see Jessop 1982, 2002b).

Some progressive states like Brazil and Colombia currently exercise well-conceived policies that enable the state to receive part of the potential ground rent from gentrification. We think that Brazil and Colombia's 'land value capture policies' are good examples to scrutinize in terms of realistic alternatives to gentrification; however, similar policies exist in Europe, and certainly in London, they have not worked (Lees 2014a). Although some may argue that the 'right to the city' is a Western European idea or even a middle-class appropriation excluding the subaltern, social movements may still find it effective for fighting gentrification and speculative urban transformation, and putting forward alternative claims. However, given planetary urbanization, it might be that the 'right to the urban' (Shin 2014a) is a better phrase, given that urbanizing societies are now beyond the administrative boundary of the 'city' (see also Souza 2015, who pushes this even further to call for the right to the *planet*). Political governance of the urban is key here.

Existing conflicts need to be contextualized in each locality, critically understood in their temporal, spatial and class dimensions, and also historicized in the ways in which rights claims have been exercised. In Montevideo, Uruguay, and Buenos Aires, Argentina, mutual-aid housing cooperatives have had a central role in the recent inner-city renewal of those cities, so that they exhibit rates of displacement which are considerably lower than in the rest of Latin America. Housing cooperatives in these cities show a way of renewal without the displacement of low income tenants (Díaz Parra and Rabasco 2013). The grassroots perform simultaneously the roles of realtors, land buyers, developers and builders, using their working-class cultural capital for community organization and even local people's construction skills.

Rather than fighting over what gentrification is and what it is not, attention needs to be paid to the actual experiences of urban change affecting urban, suburban and rural communities across the world. It is important for both intellectual endeavour and social movements to reveal the interconnectedness of the impacts of planetary urbanization, that the contemporary urbanization drive at a planetary scale is affecting all segments of populations regardless of where they live. As Shin (2014a: 515) notes:

> It will be important for the discontented to educate themselves and others to reveal the underlying logics of China's capital accumulation,

how it produces a hybrid of developmental statism and neo-liberalism, how it evades the Chinese state's own legitimacy (by constantly deviating from socialist principles and by producing prosperity at the expense of the economic hardship of the masses), and how the fate of urban inhabitants is knitted tightly with the fate of workers, villagers, and others, subject to the exploitation of the urban-oriented accumulation.

We need to unpick 'the planet's gentrified mind', and we need to be counter-cultural again, to find radical ways and insights, to operate outside social assumptions, to generate social and urban change through contestation and the presentation of realistic alternatives. Buck-Morss (2003: 11) has written about the possibilities for the Global Left in the 'expansion of the discursive field'. As McFarlane (2011: 10) has said: 'this reading of theory in travel links the circulation of theory and knowledge to possibilities for social change'. And it is to that end – social change – that this book is motivated. As Lefebvre (1996: 172) said: 'To the extent that the contours of the future city can be outlined, it could be defined by imagining the reversal of the current situation, by pushing to its limits the converted image of the world upside down.'

References

Aalbers, M. (2011) *Place, exclusion and mortgage markets*. Wiley-Blackwell, Oxford.
ACHR (Asia Coalition for Housing Rights) (1989) Evictions in Seoul, South Korea. *Environment and Urbanization* 1(1), 89–94.
Akinsami, G. (2013) Clinton: Eko Atlantic City, destination for global investment. *This Day Live* (http://www.thisdaylive.com/articles/clinton-eko-atlantic-city-destination-for-global-investment/140277/)
Alexandri, G. (2015) Unravelling the yarn of gentrification trends in the contested inner city of Athens. In Lees, L., Shin, H. B. and López-Morales, E. (eds.) *Global gentrifications: Uneven development and displacement*. Policy Press, Bristol, pp. 19–35.
Alonso, A. (1964) *Location and land use: Toward a general theory of land rent*. Harvard University Press, Cambridge.
Álvarez-Rivadulla, M. J. (2007). Golden ghettos: Gated communities and class residential segregation in Montevideo, Uruguay. *Environment and Planning A* 39(1), 47–63.
Angell, E., Hammond, T. and Van Dobben Schoon, D. (2014) Assembling Istanbul: Buildings and bodies in a world city. *City* 18(6), 644–54.
Arabindoo, P. (2011) Rhetoric of the 'slum': Rethinking urban poverty. *City* 15(6), 637–46.
Arnouts, R. and Arts, B. (2009) Environmental governance failure: The 'dark side' of an essentially optimistic concept. In Arts, B., Lagendijk, A. and van Houtum, H. (eds.) *In the disoriented state: Shifts in governmentality, territoriality and governance*. Springer, Dordrecht, pp. 201–28.
Arrighi, G. (2009) China's market economy in the long-run. In Hung, H-f. (ed.) *China and the Transformation of Global Capitalism*. Johns Hopkins University Press, Baltimore, pp. 22–49.
Arts, B., Lagendijk, A. and van Houtum, H. (2009) (eds.) *In the disoriented state: Shifts in governmentality, territoriality and governance*. Springer, Dordrecht.

REFERENCES

Ascensao, E. (2015) Slum gentrification in Lisbon, Portugal: Displacement and the imagined futures of an informal settlement. In L. Lees, H. B. Shin and E. López-Morales (eds.) *Global gentrifications: Uneven development and displacement.* Policy Press, Bristol, pp. 37–58.

Atkinson, R. (2003) Domestication by cappuccino or a revenge on urban space? Control and empowerment in the management of public space. *Urban Studies* 40(9), 1829–43.

Atkinson, R. and Bridge, G. (2005) (eds.) *Gentrification in a global context: The new urban colonialism.* Routledge, London.

Badyina, A. and Golubchikov, O. (2005) Gentrification in central Moscow – a market process or a deliberate policy? Money, power and people in housing regeneration in Ostozhenka. *Geografiska Annaler: Series B Human Geography* 87(2), 113–129.

Ballard, R. (2012) Geographies of development: Without the poor. *Progress in Human Geography* 36(5), 563–72.

Bartolome, L. J., de Wet, C., Mander, H. and Nagaraj, V. K. (2000) *Displacement, resettlement, rehabilitation, reparation and development.* Final draft prepared for the World Commission on Dams. Cape Town: World Commission on Dams.

Baviskar, A. (2007) Demolishing Delhi: World-class city in the making. In Batra, L. (ed.) *The Urban poor in globalising India: Dispossession and marginalisation,* Delhi, South Asian Dialogues on Ecological Democracy and Vasudhaiva Kutumbakam Publications.

Beauregard, R. (2003) City of superlatives. *City and Community* 2(3), 183–99.

Betancur, J. (2014) Gentrification in Latin America: Overview and critical analysis. *Urban Studies Research* DOI: 10.1155/2014/986961

Biel, R. (2000) *The new imperialism: Crisis and contradictions in North–South relations.* Zed, London.

Blanco, I. (2009) Does a 'Barcelona Model' really exist? *Local Government Studies* 35(3), 355–69.

Blanco, J., Bosoer, L. and Apaolaza, R. (2014) Gentrificación, movilidad y transporte: aproximaciones conceptuales y ejes de indagación. *Revista de Geografía Norte Grande* 58, 41–53.

Blaut, J. (1993) *The colonizer's model of the world: Geographical diffusionism and Eurocentric history.* Guilford, New York.

Boddy, M. (2007) Designer neighbourhoods: New-build residential development in non-metropolitan UK cities – the case of Bristol. *Environment and Planning A* 39(1), 86–105.

Bogaert, K. (2013) Cities without slums in Morocco? New modalities of urban governance and the *bidonville* as a neoliberal assemblage. In Samara, T., He, S.

and Chen, G. (eds.) *Right to the city in the global South: Transnational urban governance and socio-spatial transformations.* Routledge, New York, pp. 41–59.

Bondi, L. (1991) Gender divisions and gentrification: A critique. *Transactions of the Institute of British Geographers* 16(2), 190–8.

Booyens, I. (2012) Creative industries, inequality and social development: Developments, impacts and challenges in Cape Town. *Urban Forum* 23(1), 43–60.

Bose, P. S. (2014) *Urban development in India: Global Indians in the remaking of Kolkata.* Routledge, London and New York City.

____ (2013) Bourgeois environmentalism, leftist development, and neoliberal urbanism in the city of joy. In Samara, T., He, S. and Chen, G. (eds.) *Right to the city in the global South: Transnational urban governance and socio-spatial transformations.* Routledge, New York, pp. 127–51.

Brenner, N. (2009) What is critical urban theory? *City* 13(2–3), 198–207.

Brenner, N., Marcuse, P. and Mayer, M. (2012) (eds.) *Cities for people, not for profit: Critical urban theory and the right to the city.* Routledge, London.

Brenner, N., Peck, J. and Theodore, N. (2010) Variegated neoliberalization: Geographies, modalities, pathways. *Global Networks* 10(2), 1–41.

Brenner, N. and Schmid, C. (2012) Planetary urbanization. In Gandy, M. (ed.) *Urban Constellations.* Jovis, Berlin, pp. 10–13.

____ (2014) The 'urban age' in question. *International Journal of Urban and Regional Research* 38(3), 731–55.

____ (2015) Towards a new epistemology of the urban? *City* 19(2–3), 151–82.

Brenner, N. and Theodore, N. (2002) Cities and the geographies of 'actually existing neoliberalism'. *Antipode* 33(3), 349–79.

Bridge, G. (2007) A global gentrifier class? *Environment and Planning A* 39(1), 32–46.

____ (2001) Bourdieu, rational action and the time–space strategy of gentrification. *Transactions of the Institute of British Geographers* 26(2): 205–16

Bridge, G., Butler, T. and Lees, L. (2011) (eds.) *Mixed communities: Gentrification by stealth?* Policy Press, Bristol.

Briggs, L. (2002) *Reproducing empire: Race, sex, science and US imperialism in Puerto Rico.* University of California Press, Berkeley.

Brosius, C. (2010) *India's middle class: New forms of urban leisure, consumption and prosperity.* Routledge, New Delhi.

Broudehoux, A-M. (2007) Spectacular Beijing: The conspicuous construction of an Olympic metropolis. *Journal of Urban Affairs* 29(4), 383–99.

—— (2004) *The making and selling of post-Mao Beijing*. Routledge, London.
Buck-Morss, S. (2003) *Thinking past terror: Islamism and critical theory on the left*. Verso, London.
Bunnell, T., Goh, D., Lai, C. K. and Pow, C. P. (2012) Introduction: Global urban frontiers? Asian cities in theory, practice and imagination. *Urban Studies* 49(13), 2785–93.
Burdett, R. and Sudjic, D. (2011) (eds.) *Living in the endless city: The urban age project by the London School of Economics and Deutsche Bank's Alfred Herrhausen Society*. Phaidon Press, London.
Butler, T. (2010) Gentrification and globalization: The emergence of a middle range theory? (http://blogs.sciences-po.fr/recherche-villes/files/2010/01/cahier_ville_0514.pdf)
—— (2007a) For gentrification? *Environment and Planning A* 39(1), 162–81.
—— (2007b) Re-urbanizing London Docklands: Gentrification, suburbanization or new urbanism? *International Journal of Urban and Regional Research* 31(4), 759–81.
—— (2003) Living in the bubble: Gentrification and its 'others' in London. *Urban Studies* 40, 12: 2469–86.
—— (1997) *Gentrification and the middle classes*. Ashgate, Aldershot.
Butler, T. and Lees, L. (2006) Super-gentrification in Barnsbury, London: Globalisation and gentrifying global elites at the neighbourhood level. *Transactions of the Institute of British Geographers* 31, 467–87.
Cabannes, Y., Yafai, S. G., and Johnson, C. (2010) *How people face evictions*, Development and Planning Unit, University College London, London.
Cain, A. (2014) African urban fantasies: Past lessons and emerging realities. *Environment and Urbanization* 26(2), 561–67.
Calthorpe, P. (1993) *The next American metropolis: Ecology, community, and the American dream*. Princeton Architectural Press, New York.
Carbajal, N., Fernández-Dávila, T., Florez, R. and Zubiate, M. (2003) Regeneración Urbana en Guayaquil. *Medio de Construcción, Informe Especial* 173/174, Febrero/Marzo.
Cardoso, F. H. and Faletto, E. (1979) *Dependency and development in Latin America*. University of California Press, Berkeley.
Cartier, C. (2011) Neoliberalism and the neoauthoritarian city in China – Contexts and research directions. *Urban Geography* 32(8), 1110–21.
Castells, M. (1974) *La lucha de clases en Chile*. Siglo XXI, Buenos Aires.
—— (1973) Movimiento de pobladores y lucha de clases en Chile. *EURE* 3(7), 9–35.

Caulfield, J. (1994) *City form and everyday life: Toronto's gentrification and critical social practice*. University of Toronto Press, Toronto, Canada.

Cervero, R. and Murakami, J. (2009) Rail and property development in Hong Kong: Experiences and extensions. *Urban Studies* 46(10), 2019–43.

Chakrabarty, D. (2000) *Provincializing Europe: postcolonial thought and historical difference*. Princeton University Press, Princeton.

Chakravarty, S. and Qamhaieh, A. (2015) City upgraded: Redesigning and disciplining downtown Abu Dhabi. In Lees, L., Shin, H. B. and López-Morales, E. (eds.) *Global gentrifications: Uneven development and displacement*. Policy Press, Bristol, pp. 59–80.

Chang, T. C. and Teo, P. (2009) The Shophouse Hotel: Vernacular heritage in a creative city. *Urban Studies* 46(2), 341–67.

Chang, T. C. and Huang, S. (2005) Recreating place, replacing memory: Creative destruction at the Singapore River. *Asia Pacific Viewpoint* 46(3), 267–80.

Charles, S. L. (2011) Suburban gentrification: Understanding the determinants of single-family residential redevelopment, a case study of the inner-ring suburbs of Chicago, IL, 2000–10. Working paper no. W11–1. Joint Center for Housing Studies, Harvard University.

Chateau, J. and Pozo, H. (1987) Los Pobladores en el Área Metropolitana: Situación y Características. *Espacio y Poder: Los Pobladores*. Facultad Latinoamericana de Ciencias Sociales-FLACSO, Santiago, 13–71.

Chatterjee, P. (2006) *The politics of the governed: Reflections on popular politics in most of the world*. Columbia University Press, New York.

Chen, C. Y. and Webster, C. J. (2005) Homeowners' associations, collective action and the costs of private governance. *Housing Studies* 20(2), 205–20.

Chen, F. (2011) Traditional architectural forms in market oriented Chinese cities: Place for localities or symbol of culture? *Habitat International* 35(2), 410–18.

Chen, J. (2013) *A middle class without democracy: Economic growth and the prospects of democratization in China*. Oxford University Press, Oxford.

Chen, K-H. (2010) *Asia as method: Toward deimperialization*. Duke University Press.

Cheng, Z. (2012) The changing and different patterns of urban redevelopment in China: A study of three inner-city neighbourhoods. *Community Development* 43(4), 430–50.

Choi, N. (2014) Metro Manila through the gentrification lens: Disparities in urban planning and displacement risks. *Urban Studies* DOI: 10.1177/0042098014543032

Chung, H. and Unger, J. (2013) The Guangdong model of urbanisation: Collective village land and the making of a new middle class. *China Perspectives* 2013/3, 33–41.

Clark, E. (2005) The order and simplicity of gentrification – a political challenge. In Atkinson, R. and Bridge, G. (eds.) *Gentrification in a global context: The new urban colonialism.* Routledge, London, pp. 256–64.

___ (1987) *The rent gap and urban change: Case studies in Malmö 1860–1985.* Lund University Press, Lund.

Clarke, N. (2012) Actually existing comparative urbanism: Limitation and cosmopolitanism in North-South interurban partnerships. *Urban Geography* 33(6), 796–815.

Cochrane, A. (2007) *Understanding urban policy.* Blackwell, Oxford.

COHRE (2007) *Fair play for housing rights: Mega-events, Olympic Games and housing rights.* Centre on Housing Rights and Evictions, Geneva.

___ (2006) *Forced evictions: Violations of human rights, 2003–6.* Centre on Housing Rights and Evictions, Geneva.

Contreras, Y. (2011). La recuperación urbana y residencial del centro de Santiago: Nuevos habitantes, cambios socioespaciales significativos. *EURE,* 37(112), 89–113.

Cuenya, B. and Corral, M. (2011) Empresarialismo, economía del suelo y grandes proyectos urbanos: El modelo de Puerto Madero en Buenos Aires. *EURE* 37(111), 25–45.

Cummings, J. (2015) Confronting Favela Chic: the Gentrification of Informal Settlements in Rio de Janeiro, Brazil. In Lees, L., Shin, H. B. and López-Morales, E. (eds.) *Global gentrifications: Uneven development and displacement.* Policy Press, Bristol, pp. 81–99.

___ (2013) Confronting the favela chic: gentrification of informal settlements in Rio de Janeiro, Brazil. Thesis submitted to the Department of Urban Planning and Design, Harvard University Graduate School of Design.

Cunningham, S. (2009) Trojan horse or Rorschach blot? Creative industries discourse around the world. *International Journal of Cultural Policy* 15(4), 375–86.

Darling, E. (2005) The city in the country: Wilderness gentrification and the rent gap. *Environment and Planning A* 37(6), 1015–32.

Datta, A. (2015) New urban utopias of postcolonial India: 'Entrepreneurial urbanization' in Dholera smart city, Gujarat. *Dialogues in Human Geography* 5(1), 3–22.

Davidson, C. (2009) Abu Dhabi's new economy: Oil, investment and domestic development. *Middle East Policy* 16(2), 59–79.

Davidson, M. (2007) Gentrification as global habitat: A process of class formation or corporate creation? *Transactions of the Institute of British Geographers* 32, 490–506.

Davidson, M. and Lees, L. (2010) New-build gentrification: Its histories, trajectories, and critical geographies. *Population, Space and Place* 16, 395–411.

___ (2005) New-build 'gentrification' and London's riverside renaissance. *Environment and Planning A* 37, 1165–90.

Davis, L. K. (2011) International events and mass evictions: A longer view. *International Journal of Urban and Regional Research* 35(3), 582–99.

Davis, M. (2006a) Fear and money in Dubai. *New Left Review* 41, 47–68.

___ (2006b) *Planet of Slums.* Verso, London.

Dear, M. (2003) The Los Angeles School of urbanism: An intellectual history. *Urban Geography* 24(6), 493–509.

Degen, M. and García, M. (2012) The transformation of the 'Barcelona Model': An analysis of culture, urban regeneration and governance. *International Journal of Urban and Regional Research* 36(5), 1022–38.

Delgadillo, V. (forthcoming) Is Mexico City and its historic core becoming gentrified? *Urban Geography*, a special issue on Latin American gentrification, edited by López-Morales, E., Shin, H. B. and Lees, L.

___ (2014) Ciudad de México: Megaproyectos urbanos, negocios privados y resistencia social. In Hidalgo, R. and Janoschka, M. (eds.) *La ciudad neoliberal: gentrificación y exclusión en Santiago de Chile, Buenos Aires, Ciudad de Mexico y Madrid.* Universidad Católica, Santiago, pp. 199–215.

Desai, V. and Loftus, A. (2013) Speculating on slums: Infrastructural fixes in informal housing in the global South. *Antipode* 45(4), 789–808.

De Soto, H. (2000) *The Mystery of Capital: Why Capitalism Triumphs in the West and Fails Everywhere Else.* Basic Books, New York.

Díaz Parra, I. and Rabasco, P. (2013) ¿Revitalización sin gentrificación? Cooperativas de vivienda por ayuda mutua en los centros de Buenos Aires y Montevideo. *Cuadernos Geográficos* 52(2): http://revistaseug.ugr.es/index.php/cuadgeo/article/view/1516/1731

D'Monte, D. (2011) A matter of people. In Burdett, R., and Sudjic, D. (eds.) *Living in the Endless City: The Urban Age Project by the London School of Economics and Deutsche Bank's Alfred Herrhausen Society.* Phaidon Press, London, pp. 94–101.

Dolowitz, D. and Marsh, D. (2000) Learning from abroad: The role of policy transfer in contemporary policy-making. *Governance: An international journal of policy, administration, and institutions.* 13(1), 5–23.

Dos Santos Junior, O. and dos Santos, M. (2014) The right to housing, the 2014 World Cup and the 2016 Olympics: Reflections on the case of Rio de Janeiro, Brazil. In Queiroz Ribeiro, L. (ed.) *The Metropolis of Rio de Janeiro: A space in transition*. Letra Capital, Rio de Janeiro.

Doshi, S. (2015) Rethinking gentrification in India: Displacement, dispossession and the specter of development. In Lees, L., Shin, H. B. and López-Morales, E. (eds.) *Global gentrifications: Uneven development and displacement*. Policy Press, Bristol, pp. 101–19.

—— (2013) The politics of the evicted: Redevelopment, subjectivity, and difference in Mumbai's slum frontier. *Antipode* 45(4), 844–65.

Duckett, J. (1998) *The entrepreneurial state in China: Real estate and commerce departments in reform era Tianjin*. Routledge, London.

Duncan, J. and D. Ley (1982) Structural marxism and human geography: A critical assessment. *Annals of the Association of American Geographers* 72(1), 30–59.

Dutton, P. (2005) Outside the metropole: Gentrification in provincial cities or provincial gentrification? in R. Atkinson and G. Bridge (eds.) *Gentrification in a Global Context: The New Urban Colonialism* (London: Routledge) pp. 209–24.

—— (2003) Leeds calling: The influence of London on the gentrification of regional cities. *Urban Studies* 40(12), 2557–72.

Elshahed, M. (2015) The prospects of gentrification in downtown Cairo: Artists, private investment and the neglectful state. In Lees, L., Shin, H. B. and López-Morales, E. (eds.) *Global gentrifications: Uneven development and displacement*. Policy Press, Bristol, pp. 121–42.

Fang, K. and Zhang, Y. (2003) Plan and market mismatch: Urban redevelopment in Beijing during a period of transition. *Asia Pacific Viewpoint* 44(2), 149–162.

Ferguson, J. (2009) The uses of neoliberalism. *Antipode* 41(6), 166–184.

Fernandes, L. (2006) *India's new middle class: Democratic politics in an era of economic reform*. University of Minnesota Press: Minneapolis.

—— (2009) The political economy of lifestyle: Consumption, India's new middle class and state-led development. In Meier, L. and Lang, L. (eds.) *The new middle classes: Globalizing lifestyles, consumerism and environmental concern*. Spinger, pp. 219–36.

Fernández Arrigiota, M. (2010) Constructing 'the other', practicing resistance: Public housing and community politics in Puerto Rico. PhD thesis, The London School of Economics and Political Science.

Florida, R. (2002) *The rise of the creative class and how it's transforming work, leisure and everyday life*. Basic Books, New York.

Fraser, J. and Nelson, M. H. (2008) Can mixed-income housing ameliorate concentrated poverty? The significance of a geographically informed sense of community. *Geography Compass* 2(6), 2127–44.

Freeman, L. (2006) *There goes the Hood: Views of gentrification from the ground up*. Temple University Press, Philadelphia.

Fyfe, N. (2004) Zero tolerance, maximum surveillance? Deviance, difference and crime control in the late modern city. In Lees, L. (ed.) *The Emancipatory City? Paradoxes and possibilities*. Sage, London, pp. 40–56.

Gaffney, C. (forthcoming) Forging the Rings: Rio de Janeiro's pre-Olympic real-estate landscape. *Urban Geography*, special issue on Latin American gentrification, edited by E. López-Morales, H. B. Shin and L. Lees.

Garcia-Ramon, M. D. and Albet, A. (2000) Commentary: Pre-Olympics and post-Olympics Barcelona, a 'model' for urban regeneration today? *Environment and Planning A* 32, 1331–4.

Gatica, J. (1989) *Deindustrialization in Chile*. Westview Press, Boulder, San Francisco and London.

Ghertner, A. (2011) Gentrifying the state, gentrifying participation: Elite governance programs in Delhi. *International Journal of Urban and Regional Research* 35(3), 504–32.

Glass, R. (1964a) (ed.) *London: Aspects of change*. MacGibbon and Kee, London.

—— (1964b) *Urban–rural differences in Southern Asia: some aspects and methods of analysis*. UNESCO Research Centre on Social and Economic Development in Southern Asia.

Glassman, J. and Choi, Y-J (2014) The chaebol and the US military–industrial complex: Cold war geo-political economy and South Korean industrialization, *Environment and Planning A* 46(5), 1160–80.

Goetz, E. G. (2010) Desegregation in 3D: Displacement, dispersal and development in American public housing. *Housing Studies* 25(2), 137–58.

Goldman, M. (2011) Speculative urbanism and the making of the next world city. *International Journal of Urban and Regional Research* 35(3), 555–81.

Gómez, M. and González, S. (2001) A reply to Beatriz Plaza's 'The Guggenheim-Bilbao Museum effect'. *International Journal of Urban and Regional Research* 25(4), 898–900.

González, S. (2011) Bilbao and Barcelona 'in motion': How urban regeneration 'models' travel and mutate in the global flows of policy tourism. *Urban Studies* 48(7), 1397–418.

Graham, S. (2004) (ed.) *Cities, war and terrorism: Towards an urban geopolitics*. Blackwell, Cambridge.

Gramsci, A. (1971) *Selections from the Prison Notebooks*. Translated by Geoffrey Nowell Smith and Quintin Hoare. Lawrence and Wishart Limited, London.

Grimsrud, G. M. (2011) How well does the 'counter-urbanisation story' travel to other countries? The case of Norway. *Population, Space and Place* 17(5), 642–55.

Gwan'ak District Assembly (1996) *Minutes of an interim meeting: Construction committee*. Dated 28 October. Gwan'ak District Assembly, Seoul.

Ha, S-K. (2015) The endogenous dynamics of urban renewal and gentrification in Seoul. In Lees, L., Shin, H. B. and López-Morales, E. (eds.) *Global gentrifications: Uneven development and displacement*. Policy Press, Bristol, pp. 165–80.

____ (2001) Substandard settlements and joint redevelopment projects in Seoul. *Habitat International* 25, 385–97.

Hackworth, J. (2007). *The neoliberal city: Governance, ideology, and development in American Urbanism*. Cornell University Press, New York.

Hackworth, J. and Smith, N. (2001) The changing state of gentrification. *Tijdschrift poor Economische en Sociale Geografie* 92(4), 464–77.

Hajer, M. (1995) *The politics of environmental discourse: Ecological modernization and the policy process*. Clarendon Press, Oxford.

Hall, P. (1999) *Cities in civilization: Culture, innovation and urban order*. Phoenix Giant, London.

Hall, R. (2012) Housing crisis causes surge in sheds with beds. *Independent* 10 May: http://www.independent.co.uk/news/uk/home-news/housing-crisis-causes-surge-in-sheds-with-beds-7729179.html

Hamnett, C. (2003) Gentrification and the middle class remaking of inner London, 1961–2001. *Urban Studies* 40(12), 2401–26.

Harris, A. (2012) The metonymic urbanism of twenty-first-century Mumbai. *Urban Studies* 49(13), 2955–73.

____ (2008). From London to Mumbai and back again: Gentrification and public policy in comparative perspective. *Urban Studies* 45(12), 2407–28.

Hart, G. (2004) Geography and development: Critical ethnographies. *Progress in Human Geography* 28, 91–100.

Hartman, C. (1984) The right to stay put. In Geisler, C. C. and Popper, F. (eds.) *Land reform, American style*. Rowman and Allanheld, Totowa, NJ, pp. 302–18.

Hartman, C., Keating, D. and LeGates, R. (1982) *Displacement: How to fight it*. National Housing Law Project, Washington, DC.

Harvey, D. (2012) *Rebel cities: From the right to the city to the urban revolution*. Verso Books, London.

_____ (2010a) *The enigma of capital and the crises of capitalism.* Profile Books, London.

_____ (2010b) The right to the city: From capital surplus to accumulation by dispossession. In Banerjee-Guha, S. (ed.) *Accumulation by dispossession: Transformative cities in the new global order.* SAGE, London, pp. 17–32.

_____ (2007) *A brief history of neoliberalism.* Oxford University Press, Oxford.

_____ (2005) *The new imperialism.* Oxford University Press, Oxford.

_____ (1989a) *The urban experience.* Johns Hopkins University Press, Baltimore, London.

_____ (1989b) From managerialism to entrepreneurialism: The transformation of urban governance in late capitalism. *Geografiska Annaler* 71B, 3–17.

_____ (1982) *The Limits to Capital.* Blackwell, Oxford.

_____ (1978) The urban process under capitalism: A framework for analysis. *International Journal of Urban and Regional Research* 2(1–4), 101–31.

_____ (1973) *Social Justice and the City.* Edward Arnold, London.

Hasan, A. (2012) The gentrification of Karachi's coastline, unpublished paper for London Workshop 'Towards an emerging geography of gentrification in the global south' 23–4 March, London: http://arifhasan.org/wp-content/uploads/2012/08/P16_Gentrification-Karachi-Coastline.pdf

Haugaard, M. (2006) Conceptual confrontation. In Haugaard, M. and Lentner, H. (eds.) *Hegemony and power: Consensus and coercion in contemporary politics.* Lexington Books, Oxford, pp. 3–19.

Hayllar, M. R. (2010) Who owns culture and heritage? Observations on Hong Kong's experience. *International Journal of Public Policy* 5(1), 24–40.

He, S. (2012) Two waves of gentrification and emerging rights issues in Guangzhou, China. *Environment and Planning A* 44(12), 2817–33.

Healey, P., Davoudi, S., O'Toole, M., Tavsanoglu, S. and Usher, D. (1992) *Rebuilding the city: Property-led urban regeneration.* E&FN Spon, London.

Heidegger, M. (1927/1996) trans. Joan Stambaugh *Being and Time: A Translation of Sein und Zeit.* SUNY Press, Albany.

Herzer, H. (2008) (ed.) *Con el Corazon Mirando al Sur: transformaciones en el Sur de la Ciudad de Buenos Aires.* Espacio Editorial: Buenos Aires, Argentina.

Herzer, H., Di Virgilio, M. M. and Rodríguez, M. C. (2015) Gentrification in Buenos Aires: global trends and local features. In Lees, L., Shin, H. B. and López-Morales, E. (eds.) *Global gentrifications: Uneven development and displacement.* Policy Press, Bristol, pp. 199–222.

Hess, S. (2010) Nail-houses, land rights, and frames of injustice on China's protest landscape. *Asian Survey* 50(5), 908–26.

Holston, J. (2008). *Insurgent Citizenship: Disjunctions of democracy and modernity in Brazil*. Princeton University Press, Princeton and Oxford.

Hsing, Y-t. (2010) *The great urban transformation: Politics of land and property in China*. Oxford University Press, Oxford.

Hsu, J-y. and Hsu, Y-h. (2013) State transformation, policy learning, and exclusive displacement in the process of urban redevelopment in Taiwan. *Urban Geography* 34(5), 677–698.

Huang, L. (2015) Promoting private interest by public hands? The gentrification of public lands by housing policies in Taipei City. In Lees, L., Shin, H. B. and López-Morales, E. (eds.) *Global gentrifications: Uneven development and displacement*. Policy Press, Bristol, pp. 223–44.

Huang, X. and Yang, Y. (2010) The characteristics and forming mechanisms of gentrification in cities of Western China: The case study in Chengdu city (in Chinese). *Progress in Geography* (dili kexue jinzhan) 29(12), 1532–40.

Hyra, D. (2008) *The new urban renewal: The economic transformation of Harlem and Bronzeville*, University of Chicago Press, Chicago.

Innes, M. (1999) An iron fist in an iron glove? The zero tolerance policing debate. *The Howard Journal* 38, 397–410.

International Alliance of Inhabitants (2008) Vigilias por el derecho a vivir en el centro histórico de Lima. 2 October: http://esp.habitants.org/campana_cero_desalojos/

Inzulza-Contardo, J. (2012) 'Latino gentrification'?: focusing on physical and socioeconomic patterns of change in Latin American inner cities. *Urban Studies* 49(10), 2085–107.

Iossifova, D. (2009) Negotiating livelihoods in a city of difference: narratives of gentrification in Shanghai. *Critical Planning* 16(2), 99–116.

İslam, T. and Sakızlıoğlu, B. (2015) The making of, and resistance to, state-led gentrification in Istanbul, Turkey. In Lees, L., Shin, H. B. and López-Morales, E. (eds.) *Global gentrifications: Uneven development and displacement*. Policy Press, Bristol, pp. 245–64.

Jacobs, J. (1961) *The death and life of great American cities*. Random House, New York.

Janoschka, M., Sequera, J. and Salinas, L. (2014) Gentrification in Spain and Latin America – a critical dialogue. *International Journal of Urban and Regional Research* DOI: 10.1111/1468-27.12030

Jaramillo, S. (2008). *Hacia una teoría de la renta de suelo urbano*. Uniandes, Bogotá.

Jazeel, T. and McFarlane, C. (2010) The limits of responsibility: A postcolonial politics of academic knowledge production. *Transactions, Institute of British Geographers* 35(1), 109–24.

Jessop, B. (2002a) Liberalism, neoliberalism, and urban governance: A state theoretical perspective. *Antipode* 34, 452–72.

―――― (2002b) *The future of the capitalist state.* Polity, Cambridge.

―――― (1982) *The capitalist state.* Martin Robertson and Company, Oxford.

Joseph, M. (2006) Is mixed-income development an antidote to urban poverty? *Housing Policy Debate*, 17(2), 209–34.

Jou, S-C., Clark, E. and Chen, H-W. (2014) Gentrification and revanchist urbanism in Taipei? *Urban Studies*: http://dx.doi.org/10.1177/0042098014541970

Kaviraj, S. (1997) Filth and the public sphere: Concepts and practices about space in Calcutta. *Public Culture* 10(1), 83–113.

Keil, R. (2013) (ed.) *Suburban constellations: Governance, land and infrastructure in the 21st century.* Jovis, Berlin.

Kelling, G. L. and Wilson, J. Q. (1982) Broken windows: The police and neighborhood safety. *Atlantic Monthly* http://www.theatlantic.com/magazine/archive/1982/03/broken-windows/304465/

Kharas, H. and Goertz, G. (2010) The new global middle class: A cross-over from West to East. In Li, C. (ed.) *China's emerging middle class: Beyond economic transformation.* Brookings Institution Press, Washington, D. C.

Kim, H-h. (1998) South Korea: Experiences of eviction in Seoul. In Azuela, A., Duhau, E. and Ortiz, E. (eds.) *Evictions and the right to housing: Experience from Canada, Chile, the Dominican Republic, South Africa and South Korea.* International Development Research Centre, Canada, Ottawa, Ont., pp. 199–232.

Kim, J. (2013) *Chinese labor in a Korean factory: Class, ethnicity and productivity on the shop floor in globalising China.* Stanford University Press, Stanford.

Kim, K. and Nam, Y-W. (1998) Gentrification: Research trends and arguments. *Journal of Korea Planning Association* 33(5), 8–97.

Kingsley, G. T., Johnson, J. and Pettit, K. L. S. (2003) Patterns of Section 8 relocation in the HOPE VI program. *Journal of Urban Affairs* 25(4), 427–47.

Kong, L. (2012) Ambitions of a global city: Arts, culture and creative economy in post-crisis Singapore. *International Journal of Cultural Policy* 18(3), 279–94.

Kong, L. and O'Connor, J. (2009) *Creative economies, creative cities: Asian-European perspectives.* Springer, Dordrecht.

Kong, L. and Yeoh, B. S. A. (1994) Urban conservation in Singapore: A survey of state policies and popular attitudes. *Urban Studies* 31(2), 247–65.

Koo, H. (1991) Middle classes, democratization, and class formation: The case of South Korea. *Theory and Society* 20(4), 485–509.

Krijnen, M. and De Beukelaer, C. (2015) Capital, state and conflict: The various drivers of diverse gentrification processes in Beirut, Lebanon. In Lees, L., Shin, H. B. and López-Morales, E. (eds.) *Global gentrifications: Uneven development and displacement.* Policy Press, Bristol, pp. 285–309.

La Grange, A. and Pretorius, F. (2014) State-led gentrification in Hong Kong. *Urban Studies* DOI: 10.1177/0042098013513645

Larner, W. (2000) Neo-liberalism: Policy, ideology, governmentality. *Studies in Political Economy* 63, 5–26.

Larner, W. and Laurie, N. (2010) Travelling technocrats, embodied knowledges: Globalising privatisation in telecoms and water. *Geoforum* 41(2), 218–26.

Lee, C. K. (2008) Rights activism in China. *Contexts* 7(3), 14–19.

Lee, L. (2001) Shanghai modern: Reflections on urban culture in China in the 1930s. In Gaonkar, D. (ed.) *Alternative modernities.* Duke University Press, Durham, pp. 86–122.

Lees, L. (forthcoming) Doing comparative urbanism in gentrification studies: Fashion or progress? In Silver, H. (ed.) *Comparative urban studies.* Routledge, New York.

—— (2014a) Gentrification in the global South? In Parnell, S. and Oldfield, S. (eds.) *The Routledge handbook on cities of the global South.* Routledge, pp. 506–21.

—— (2014b) The urban injustices of New Labour's 'new urban renewal': The case of the Aylesbury Estate in London. *Antipode* 46(4), 921–47.

—— (2014c) The death of sustainable communities in London? In Imrie, R. and Lees, L. (eds.) *Sustainable London? The future of a global city.* Policy Press, Bristol, pp. 149–72.

—— (2012) The geography of gentrification: Thinking through comparative urbanism. *Progress in Human Geography* 36(2), 155–71.

—— (2008) Gentrification and social mixing: Towards an inclusive urban renaissance? *Urban Studies* 45(12), 2449–70.

—— (2004) (ed.) *The Emancipatory City: Paradoxes and possibilities?* Sage, London.

—— (2003) Super-gentrification: The case of Brooklyn Heights, New York City. *Urban Studies* 40(12), 2487–509.

—— (2000) A re-appraisal of gentrification: Towards a geography of gentrification. *Progress in Human Geography* 24, 389–408.

Lees, L. and Ley, D. (2008) Introduction: Gentrification and public policy. *Urban Studies* 45(12), 2379–84.

Lees, L., Shin, H. B. and López-Morales, E. (eds.) (2015) *Global gentrifications: Uneven development and disparity.* Policy Press, Bristol.

Lees, L., Slater, T. and Wyly, E. (2010) *The Gentrification Reader.* Routledge, London.

___ (2008) *Gentrification.* Routledge, New York.

Lefebvre, H. (2003) *The urban revolution.* Translated by Robert Bononno from the original French publication, La Révolution urbaine (1970 edn). University of Minnesota Press, Minneapolis, MN.

___ (1996) *Writings on Cities.* Translated and edited by E. Kofman and E. Lebas. Blackwell, Oxford.

Lemanski, C. (2014) Hybrid gentrification in South Africa: Theorising across southern and northern cities. *Urban Studies* 51(14), 2943–60.

Lemanski, C. and Lama-Rewal, S. T. (2013) The 'missing middle': Class and urban governance in Delhi's unauthorized colonies. *Transactions of the Institute of British Geographers* 38(1), 91–105.

Lett, D. (1998) *In Pursuit of status: The making of South Korea's new urban middle class.* Harvard University Press.

Levien, M. (2011) Special economic zones and accumulation by dispossession in India. *Journal of Agrarian Change* 11(4), 454–83.

Ley, D. (2011) Social mix in liberal and neoliberal times: Social mixing and the historical geography of gentrification. In Bridge, G., Butler, T. and Lees, L. (eds.) *Mixed communities: Gentrification by stealth?* Policy Press, Bristol, pp. 53–68.

___ (2003) Artists, aestheticisation and the field of gentrification. *Urban Studies* 40(12), 2527–44.

___ (1996) *The new middle class and the remaking of the central city.* Oxford University Press, Oxford.

___ (1994) Gentrification and the politics of the new middle class. *Environment and Planning D: Society and Space* 12, 53–74.

___ (1987) Styles of the times: Liberal and neo-conservative landscapes in inner Vancouver, 1968–986. *Journal of Historical Geography* 13(1), 40–56.

Ley, D. and Dobson, C. (2008) Are there limits to gentrification? The contexts of impeded gentrification in Vancouver. *Urban Studies* 45(12), 2471–98.

Ley, D. and Teo, S. Y. (2014) Gentrification in Hong Kong? Epistemology vs. ontology. *International Journal of Urban and Regional Research* 38(4), 1286–303.

Li, L. and O'Brien K. J. (2008) Protest leadership in rural China. *The China Quarterly* 193, 1–23.

Lim, H., Kim, J., Potter, C. and Bae, W. (2013) Urban regeneration and gentrification: Land use impacts of the Cheonggye Stream restoration project on the Seoul's central business district. *Habitat International* 39, 192–200.

London Tenants Federation, Lees, L., Just Space, and Southwark Notes Archive Group. (2014) *An anti-gentrification handbook for council estates in London*. London.
López-Morales, E. (forthcoming) Gentrification in Santiago, Chile: A property-led process of dispossession and exclusion. *Urban Geography*.
____ (2013a) Gentrificación en Chile: aportes conceptuales y evidencias para una discusión necesaria. *Revista de Geografía Norte Grande* 56, 31–52.
____ (2013b) *Urbanismo proempresarial y destrucción creativa: un estudio de caso de la estrategia de renovación urbana en el pericentro de Santiago de Chile, 1990–2005*. Infonavit-Redalyc, México City.
____ (2013c) Insurgency and institutionalized social participation in local-level urban planning: The case of PAC comuna, Santiago de Chile, 2003–5. In Samara, T., He, S. and Chen, G. (eds.) *Locating right to the city in the global South: Transnational urban governance and socio-spatial transformations*. Routledge, New York, pp. 221–46.
____ (2011) Gentrification by ground rent dispossession: The shadows cast by large scale urban renewal in Santiago de Chile. *International Journal of Urban and Regional Research* 35(2), 330–57.
____ (2010) Real estate market, state-entrepreneurialism, and urban policy in the 'gentrification by ground rent dispossession' of Santiago de Chile. *Journal of Latin American Geography* 9(1), 145–73.
López-Morales, E., Gasic, I., and Meza, D. (2012) Urbanismo Pro-Empresarial en Chile: Políticas y planificación de la producción residencial en altura en el pericentro del Gran Santiago. *Revista INVI*, 28(76), 75–114.
López-Morales, E., Meza, D., and Gasic, I. (2014) Neoliberalismo, regulación ad-hoc de suelo y gentrificación: El historial de la renovación urbana del sector Santa Isabel, Santiago. *Norte Grande* (58), 161–77.
López-Morales, E. and Ocaranza, M. (2012) La Victoria de Pedro Aguirre Cerda: Ideas para una renovación urbana sin gentrificación para Santiago. *Revista de Urbanismo* 27, 42–63.
Luckman, S., Gibson, C., and Lea, T. (2009) Mosquitoes in the mix: How transferable is creative city thinking?, *Singapore Journal of Tropical Geography* 30(1), 7–85.
Ma, L. and Wu, F. (2005) (eds.) *Restructuring the Chinese city: Changing society, economy and space*. Routledge, London.
McCann, E. J. (2004) 'Best Places': Inter-urban competition, quality of life, and popular media discourse. *Urban Studies* 41(10), 1909–29.
____ (2008) Expertise, truth, and urban policy mobilities: Global circuits of knowledge in the development of Vancouver, Canada's 'Four Pillar' Drug Strategy. *Environment and Planning A* 40(4), 885–904.

McCann, E. and Ward, K. (2011) (eds.) *Mobile urbanism: Cities and policymaking in the global age.* University of Minnesota Press, Minneapolis.

___ (2010) Relationality/territoriality: Toward a conceptualization of cities in the world. *Geoforum* 41(4), 175–184.

McFarlane, C. (2011) *Learning the city: Knowledge and translocal assemblage.* Blackwell Publishing.

___ (2010) The comparative city: Knowledge, learning, urbanism. *International Journal of Urban and Regional Research* 34(4), 725–42.

___ (2006) Crossing borders: Development, learning, and the North–South divide. *Third World Quarterly* 27(8), 1413–437.

Maloutas, T. (2011) Contextual diversity in gentrification research. *Critical Sociology* 38(1), 33–48.

Marcuse, P. (1986) Abandonment, gentrification and displacement: The linkages in New York City. In Smith, N. and Williams, P. (eds.) *Gentrification of the city.* Unwin Hyman, London, pp. 153–77.

___ (1985a) Gentrification, abandonment and displacement: Connections, causes and policy responses in New York City. *Journal of Urban and Contemporary Law* 28, 195–240.

___ (1985b) To control gentrification: Anti-displacement zoning and planning for table residential districts. *New York University Review of Law and Social Change* 13, 931–52.

Marshall, T. (2004) (ed.) *Transforming Barcelona.* Routledge, London.

___ (2000) Urban planning and governance: Is there a Barcelona model? *International Planning Studies* 5(3), 299–319.

Marx, K. (1973) *Grundrisse: Foundations of the critique of political economy.* Penguin Books Ltd, Middlesex, England.

Marx, K. and Engels, F. (1848/1967) *The Communist Manifesto.* Penguin Books, London.

Massey, D. (2011) A counterhegemonic relationality of place. In McCann, E. and Ward, K. (eds.) *Mobile urbanism: Cities and policymaking in the global age.* University of Minnesota Press, Minneapolis, pp. 1–14.

___ (2007) *World city.* Polity, Cambridge.

___ (1993) Power-geometry and a progressive sense of place. In Bird, J., Curtis, B., Putnam, T., Robertson, G. and Tickner, L. (eds.) *Mapping the futures: Local cultures, global change.* Routledge, London, pp. 60–70.

___ (2004) Geographies of responsibility, *Geografiska Annaler* 86B, 5–18.

Mayer, M. (2003) The onward sweep of social capital: Causes and consequences for understanding cities, communities and urban movements. *International Journal of Urban and Regional Research* 27(1), 110–32.

Mbembe, A. and Nuttall, S. (2008) 'Introduction: Afropolis', in S. Nuttall and A. Mbembe (eds.) *Johannesburg: The elusive metropolis*. Duke University Press, pp. 1–33.

Mehta, L. (2009) (ed.) *Displaced by development: Confronting marginalisation and gender injustice*. Sage, New Delhi.

Merrifield, A. (2014) *The new urban question*. Pluto Press, London.

____ (2013a) The urban question under planetary urbanization. *International Journal of Urban and Regional Research* 37(3), 909–22.

____ (2013b) *The politics of the encounter: Urban theory and protest under planetary urbanization*. University of Georgia Press.

____ (2011) *Magical marxism: Subversive politics and the imagination*. Pluto Press, London.

Mobrand, E. (2008) Struggles over unlicensed housing in Seoul, 1960–80. *Urban Studies* 45(2), 367–89.

Monclús, F. J. (2003) The Barcelona model: An original formula? From 'reconstruction' to strategic urban projects (1979–2004). *Planning Perspectives* 18, 399–421.

Moore, S. (2013) What's wrong with best practice? Questioning the typification of new urbanism. *Urban Studies* 50(11), 2371–87.

Moreno, L. (2013) The urban process under financialised capitalism. *City* 18(3), 244–68.

Morales-Cruz, M. (2012) Lawyers and 'social' movements: A story about the Puerto Rico 'Zero Evictions' coalition. http://www.law.yale.edu/documents/pdf/sela/SELA12_Morales-Cruz_CV_Eng_20120508.pdf

Mountz, A. and Curran, W. (2009) Policing in drag: Giuliani goes global with the illusion of control. *Geoforum* 40(6), 1033–40.

MovingCities (2012) Making Creative City/Urbanus workshop. 5 September: http://movingcities.org/movingmemos/making-creative-city-urbanus-workshop/

Mueller, G. (2014) Liberalism and Gentrification. *Jacobin*, 26 September, https://www.jacobinmag.com/2014/09/liberalism-and-gentrification

Municipal Dreams (2014) The Aylesbury estate, Southwark: 'State-led gentrification'? Blog entry on 14 January: http://municipaldreams.wordpress.com/2014/01/14/the-aylesbury-estate-southwark-state-led-gentrification/

Murray, M. J. (2009) Fire and ice: Unnatural disasters and the disposable urban poor in post-Apartheid Johannesburg. *International Journal of Urban and Regional Research* 33(1), 165–92.

Myers, G. (2014) From expected to unexpected comparisons: Changing the flows of ideas about cities in a postcolonial urban world. *Singapore Journal of Tropical Geography* 35(1), 104–18.

News Guangdong (2007) Shenzhen Launches OCT-LOFT. 8 February: http://www.newsgd.com/culture/culturenews/200702080044.htm

Nijman, J. (2007) Introduction–comparative urbanism. *Urban Geography* 28(1), 1–6.

Nobre, E. (2003) Urban regeneration experiences in Brazil: Historical perspectives, tourism development and gentrification in Salvador da Bahia. *Urban Design International* 7(2), 109–24.

O'Connor, J. (2009) Shanghai moderne: Creative economy in a creative city? In Kong, L. and O'Connor, J. (eds.) *Creative economies/creative cities: Asian–European perspectives*. Springer, pp. 174–91.

O'Connor, J. and Liu, L. (2014) Shenzhen's OCT-LOFT: Creative space in the city of design. *City, Culture and Society* 5, 131–8.

Onatu, G. (2010) Mixed-income housing development strategy: Perspective on Cosmo City, Johannesburg, South Africa. *International Journal of Housing Markets and Analysis* 3(3), 203–15.

Ong, A. (2006) *Neoliberalism as an exception: Mutations of citizenship and sovereignty*. Duke University Press, Durham, NC.

Osman, S. (2011) *The invention of Brownstone Brooklyn: Gentrification and the search for authenticity in postwar New York*. Oxford University Press, Oxford.

Park, B-G. (1998) Where do tigers sleep at night? The state's role in housing policy in South Korea and Singapore. *Economic Geography* 74(3), 272–88.

Park, B-G., Hill, R. C. and Saito, A. (2012) (eds.) *Locating neoliberalism in East Asia: Neoliberalizing spaces in developmental states*. Wiley-Blackwell, Chichester.

Parnell, S. (1997) South African cities: Perspectives from the ivory tower of urban studies. *Urban Studies* 34(5–6), 891–906.

Parnreiter, C. (2011) Strategic urban planning: Towards the making of a transnational urban policy? *GaWC Research Bulletin* 385. http://www.lboro.ac.uk/gawc/rb/rb385.html

Peck, J. (2010) *Constructions of neoliberal reason*, Oxford University Press, Oxford; New York.

—— (2005) Struggling with the creative class. *International Journal of Urban and Regional Studies* 29(4), 740–70.

—— (2003) Geography and public policy: Mapping the penal state, *Progress in Human Geography* 27, 222–32.

—— (2002) Political economies of scale: Fast policy, interscalar relations, and neoliberal workfare. *Economic Geography* 78(3), 331–60.

—— (2001) Neoliberalizing states: Thin policies/hard outcomes. *Progress in Human Geography* 25(3), 445–55.

Peck, J. and Theodore, N. (2010a) Mobilizing policy: Models, methods and mutations. *Geoforum* 41, 169–74.
____ (2010b) Recombinant workfare, across the Americas: Transnationalizing fast welfare policy. *Geoforum* 41(2), 195–208.
People's Daily (2009) 'Heaven Street': 'Fake-over' kills business. 18 June: http://en.people.cn/90001/90782/6681213.html
Phillips, M. (2004) Other geographies of gentrification. *Progress in Human Geography* 28(1), 5–30.
____ (2002) The production, symbolization and socialization of gentrification: Impressions from two Berkshire villages. *Transactions of the Institute of British Geographers* 27(3), 282–308.
____ (1993) Rural gentrification and the process of class colonization. *Journal of Rural Studies* 9(2), 123–40.
Porter, L. (2009) Planning displacement: The real legacy of major sporting events. *Planning Theory and Practice* 10(3), 395–418.
Porteous, J. D. and Smith, S. E. (2001) *Domicide: The global destruction of home*. McGill-Queen's University Press.
Potts, D. (2011) Shanties, slums, breeze blocks and bricks: (Mis)understandings about informal housing demolitions in Zimbabwe. *City* 15(6), 709–21.
Poulantzas, N, (1975) *Classes in contemporary capitalism*. New Left Books, London.
Pradilla, E. (2008) Centros comerciales, terciarización, y privatización de lo público en la Zona Metropolitana del Valle de México. *Ciudades*, 79.
Pratt, A. (2009) Policy transfer and the field of the cultural and creative industries: What can be learned from Europe? In Kong, L. and O'Connor, J. (eds.) *Creative economies, creative cities*. Springer, Dordrecht, pp. 9–23.
Préteceille, E. (2007) Is gentrification a useful paradigm to analyse social changes in the Paris metropolis? *Environment and Planning A* 39(1), 10–31.
Prince, R. (2010) Globalising the creative industries concept: Travelling policy and transnational policy communities. *Journal of Law, Arts Management and Society* 40(2), 119–39.
Qiu, J. H. (2002) Consideration of 'gentrification' in contemporary Chinese renewal (in Chinese). *Tropical Geography* (redai dili) 22(2), 125–9.
Queiroz Ribeiro, L. (2013) *Transformações na Ordem Urbana das Metrópoles Brasileiras: 1980/2010. Hipóteses e estratégia teórico-metodológica para estudo comparativo*. Observatório das Metrópoles.
Queiroz Ribeiro, L. and dos Santos Junior, O. (2007) *As metrópoles e a questao social brasileira*. Revan, Rio de Janeiro.

Quijano, A. (1968). Dependencia, cambio social y urbanizacion en Latinoamerica. *Revista Mexicana de Sociología* 30(3), 525–70.

Raco, M. (2012) The privatisation of urban development and the London Olympics 2012. *City* 16(4), 452–60.

Rai, S. (1995) Gender in China. In Benewick, R. and Wingrove, P. (eds.) *China in the 1990s*. Macmillan Press, London, pp. 181–92.

Rao, V. (2006) Slum as theory: The South/Asian city and globalization. *International Journal of Urban and Regional Research* 30, 225–32.

Ren, X. (2013) *Urban China*. Polity, Cambridge.

___ (2008) Forward to the past: Historical preservation in globalizing Shanghai. *City and Community* 7(1), 23–43.

Ren, X. and Weinstein, L. (2013) Urban governance, mega-projects and scalar transformations in China and India. In T. Samara, S. He and G. Chen (eds.) *Right to the city in the global South: Transnational urban governance and sociospatial transformations*. Routledge, New York, pp. 107–26.

Rerat, P. and Lees, L. (2011) Spatial capital, gentrification and mobility: Evidence from Swiss core cities. *Transactions of the Institute of British Geographers* 36, 126–42.

Revkin, A. C. (2009) Peeling back pavement to expose watery havens. *The New York Times* 16 July: http://www.nytimes.com/2009/07/17/world/asia/17daylight.html

Richardson, T. and Jensen, O. B. (2003) Linking discourse and space: Towards a cultural sociology of space in analysing spatial policy discourses. *Urban Studies* 40(1), 7–22.

Robinson, J. (2011a) Cities in a world of cities: The comparative gesture. *International Journal of Urban and Regional Research* 35(1), 1–23.

___ (2011b) The spaces of circulating knowledge: City strategies and global urban governmentality. In McCann, E. and Ward, K. (eds.) *Mobile urbanism: City policy-making in the global age*. University of Minnesota Press, Minneapolis, pp. 15–40.

___ (2011c) Comparisons: Colonial or cosmopolitan? *Singapore Journal of Tropical Geography* 32(2), 125–40.

___ (2006) *Ordinary cities: Between modernity and development*. Routledge, London.

___ (2003) Postcolonialising geography: Tactics and pitfalls. *Singapore Journal of Tropical Geography* 24, 273–89.

___ (2002) Global and world cities: A view off the map. *International Journal of Urban and Regional Research* 26, 531–54.

Rodan, G. (1992) Singapore: Emerging tensions in the 'dictatorship of the middle class'. *The Pacific Review* 5(4), 370–81.

Rodriguez, A., Martinez, E. and Guenaga, G. (2001) Uneven redevelopment: New urban policies and socio-spatial fragmentation in metropolitan Bilbao. *European Urban and Regional Studies* 8, 161–78.

Rofe, M. M. (2003) 'I want to be global': Theorising the gentrifying class as an emergent elite global community. *Urban Studies* 40(12), 2511–26.

Rojas, E. (2004). *Volver al Centro. La recuperación de áreas urbanas centrales.* BID, Washington DC.

Roller, Z. (2011) Vila Autódromo favela resists eviction in Rio. *The Rio Times*, 11 October: http://riotimesonline.com/brazil-news/rio-politics/vila-autodromo-favela-resists-eviction-in-rio

Rose, D. (1984) Rethinking gentrification: Beyond the uneven development of marxist urban theory. *Environment and Planning D: Society and Space* 2, 47–74.

Ross, A. (2007) Nice work if you can get it: The mercurial career of creative industries policy. In Lovink, G. and Rossiter, N. (eds.) *My creativity reader: A critique of creative industries.* Institute of Network Cultures, Amsterdam, pp. 17–40.

Roy, A. (2009) The 21st century metropolis: New geographies of theory. *Regional Studies* 43, 819–30.

____ (2005) Urban informality: toward an epistemology of planning. *Journal of the American Planning Association*, 71(2), 147–58.

____ (2003) Paradigms of propertied citizenship: Transactional techniques of analysis. *Urban Affairs Review* 38, 463–91.

Roy, A. and Ong, A. (2011) (eds.) *Worlding cities: Asian experiments and the art of being global.* Wiley-Blackwell.

Ruiz-Tagle, J. (2014) Bringing inequality closer: A comparative urban sociology of socially diverse neighborhoods. Unpublished PhD thesis, Illinois University of Chicago.

Sakızlıoğlu, N. B. and Uitermark, J. (2014) The symbolic politics of gentrification: The restructuring of stigmatized neighborhoods in Amsterdam and Istanbul. *Environment and Planning A* 46, 1369–85.

Salomón, M. (2009) Local governments as foreign policy actors and global cities network-makers: The cases of Barcelona and Porto Alegre. *GaWC Research Bulletin* 305. http://www.lboro.ac.uk/gawc/rb/rb305.html

Samara, T., He, S. and Chen, G. (2013) (eds.) *Right to the city in the Global South: Transnational urban governance and socio-spatial transformations.* Routledge, New York.

Sandroni, P. (2011) Recent experience with land value capture in São Paulo, Brazil. *Land Lines* 23(3), 14–19.

Sangal, S., Nagrath, S. and Singla, G. (2010) The alternative urban futures report – urbanization and sustainability in India: An interdependent agenda. Mirabilis Advisory (see http://assets.wwfindia.org/downloads/urbanisation_report.pdf)

Sassen, S. (2006) *Cities in a world economy*. Pine Forge Press.

Satiroglu, I. and Choi, N. (2015) (eds.) *Development-induced displacement and resettlement: New perspectives on persisting problems*. Routledge, Oxford.

Schill, M. and Nathan, R. (1983) *Revitalizing America's cities: Neighborhood reinvestment and displacement*. State University of New York Press, Albany.

Schindler, S. (2015) Governing the twenty-first century metropolis and transforming territory. *Territory, Politics, Governance* 3(1), 7–26.

Schulman, S. (2012) *The gentrification of the mind: Witness to a lost imagination*. University of California Press.

Schumpeter, J. A. (1976) *Capitalism, Socialism and Democracy*. Routledge, London.

Sekularac, I. (2015) Serbia seals deal with Abu Dhabi developer for controversial Belgrade makeover. *Reuters* 26 April: http://uk.reuters.com/article/2015/04/26/uk-serbia-emirates-belgrade-idUKKBN0NH0I920150426

Sengupta, U. (2013) Inclusive development? A state-led land development model in New Town, Kolkata. *Environment and Planning C: Government and Policy* 31(2), 357–76.

Sennett, R. (1998) *The corrosion of character: The personal consequences of work in the new capitalism*. W. W. Norton and Company, New York.

Seo, J. and Chung, S. (2012) Impact of entrepreneurship in the public sector: Cheonggye Stream Restoration Project in the Seoul metropolitan city. *Asia Pacific Journal of Public Administration* 34(1), 71–93.

Seoul Municipal Government (1991) Research report on housing policy for the urban poor in Seoul (in Korean: Seoul-si Jeosodeugcheung-ui Jutaegjeongchaeg-e Gwanhan Yeongu Bogo). Seoul Municipal Government, Seoul.

Shao, Q. (2013) *Shanghai gone: Domicide and defiance in a Chinese megacity*. Rowman and Littlefield, Lanham, MD.

Shaw, W. (2011) Gentrification without social mixing in the rapidly urbanising world of Australasia. In Bridge, G., Butler, T. and Lees, L. (eds.) *Mixed communities: Gentrification by stealth?* Policy Press, Bristol, pp. 43–52.

Sheppard, E. and Leitner, H. (2010) Quo vadis neoliberalism? The remaking of global capitalist governance after the Washington Consensus. *Geoforum* 41(2), 185–94.

Shih, M. (2010) The evolving law of disputed relocation: Constructing inner-city renewal practices in Shanghai, 1990–2005. *International Journal of Urban and Regional Research* 34(2), 350–64.

Shin, H. B. (2015) Urbanization in China, In Wright, J. (ed.) *International Encyclopedia of Social and Behavioral Sciences* (2nd edn). Elsevier, pp. 973–9.

___ (2014a) Contesting speculative urbanisation and strategising discontents. *City* 18(4–5), 509–16.

___ (2014b) Urban spatial restructuring, event-led development and scalar politics. *Urban Studies* 51(14), 2961–78.

___ (2013) The right to the city and critical reflections on China's property rights activism. *Antipode* 45(5), 1167–89.

___ (2012) Unequal cities of spectacle and mega-events in China. *City* 16(6), 728–44.

___ (2010) Urban conservation and revalorisation of dilapidated historic quarters: The case of Nanluoguxiang in Beijing. *Cities* 27 (Supplement 1), S43–S54.

___ (2009a) Property-based redevelopment and gentrification: The case of Seoul, South Korea. *Geoforum* 40(5), 906–17.

___ (2009b) Residential redevelopment and entrepreneurial local state: The implications of Beijing's shifting emphasis on urban redevelopment policies. *Urban Studies* 46(13), 2815–39.

___ (2009c) Life in the shadow of mega-events: Beijing Summer Olympiad and its impact on housing. *Journal of Asian Public Policy* 2(2), 122–41.

___ (2008) Living on the edge: Financing post-displacement housing in urban redevelopment projects in Seoul. *Environment and Urbanization* 20(2), 411–26.

___ (2006) Transforming urban neighbourhoods: limits of developer-led partnership and benefit-sharing in residential redevelopment, with reference to Seoul and Beijing. PhD thesis, The London School of Economics and Political Science.

Shin, H. B. and Kim, S-H. (2015) The developmental state, speculative urbanisation and the politics of displacement in gentrifying Seoul. *Urban Studies* DOI: 10.1177/0042098014565745

Shin, H. B. and Li, B. (2013) Whose Games? The costs of being 'Olympic citizens' in Beijing. *Environment and Urbanization* 25(2), 549–66.

Shin, H., Lees, L. and López-Morales, E. [guest editors] (special issue) (2016) Locating Gentrification in East Asia, *Urban Studies*.

Sidaway, J., Woon, C. Y. and Jacobs, J. (2014) Planetary postcolonialism. *Singapore Journal of Tropical Geography* 35(1): 4–21.

Sigler, T. and Wachsmuth, D. (2015) Transnational gentrification: Globalisation and neighbourhoood change in Panama's Casco Antiguo. *Urban Studies* DOI: 10.1177/0042098014568070

Simone, A. (2004) *For the city yet to come: Changing life in four African cities.* Duke University Press, Durham, NC.

Singerman, D. and Amar, P. (2006) (eds.) *Cairo Cosmopolitan: politics, culture and urban space in the new Middle East.* AUC Press, Cairo.

Slater, D. (2010) *Ordering power: Contentious politics and authoritarian Leviathans in Southeast Asia.* Cambridge University Press.

Slater, T. (2015) Planetary rent gaps. *Antipode* DOI: 10.1111/antl.12185.

___ (2010) Still missing Marcuse: Hamnett's foggy analysis in London town. *City* 14(1), 170–79.

___ (2006) The eviction of critical perspectives from gentrification research. *International Journal of Urban and Regional Research* 30(4), 737–57.

Smith, D. P. (2005) 'Studentification': The gentrification factory? In Atkinson, R. and Bridge, G. (eds.) *Gentrification in a global context: The new urban colonialism.* Routledge, Oxford, pp. 72–89.

Smith, N. (2008) *Uneven development: Nature, capital and the production of space* (3rd edn). University of Georgia Press.

___ (2002) New globalism, new urbanism: Gentrification as global urban strategy. *Antipode* 34(3), 427–50.

___ (2000) Gentrification. In Johnston, R. J., Gregory, D., Pratt, G. and Watts, M. (eds.) *The dictionary of human geography* (4th edn). Blackwell, Oxford, pp. 294–5.

___ (1996) *The new urban frontier: Gentrification and the revanchist city.* London and New York: Routledge.

___ (1982) Gentrification and uneven development. *Economic Geography* 58(2), 139–55.

___ (1979) Toward a theory of gentrification: A back to the city movement by capital not people. *Journal of the American Planning Association* 45, 538–48.

Smolka, M. (2013) *Implementing value capture in Latin America: Policies and tools for urban development.* Policy Focus Report Series. Lincoln Institute of Land Policy, Cambridge, MA.

Souza, M.L.De. (2015) From the 'right to the city' to the right to the *planet. City* 19(4), 408–43.

Spivak, G. (2003) *Death of a discipline.* Columbia University Press, New York.

___ (1993) *Outside in the teaching machine.* Routledge, London.

___ (1985) Three women: Texts and a critique of imperialism. *Critical Inquiry* 12, 43–61.

Steinberg, F. (2001) *Planificación Estratégica Urbana en América Latina: Experiencias de Construcción y Gestión del Futuro.* SINPA, Santa Cruz de la Sierra, Bolivia.

Swanson, K. (2010) *Begging as a path to progress: Indigenous women and children and the struggle for Ecuador's urban spaces.* University of Georgia Press, Atlanta.

___ (2007) Revanchist urbanism heads south: The regulation of indigenous beggars and street vendors in Ecuador. *Antipode* 39(4), 708–28.

Sýkora, L. (1996) Economic and social restructuring and gentrification in Prague. *Geographica* 37, 71–81.

___ (1993) City in transition: The role of the rent gap in Prague's revitalization. *Tijdschrift voor Economisce en Sociale Geografie* 84(4), 281–93.

Tang, W-S. (2008) Hong Kong under Chinese sovereignty: Social development and a land (re)development regime. *Eurasian Geography and Economics* 49(3), 341–61.

Tang, W-S. and Chung, H. (2002) Rural–urban transition in China: Illegal land use and construction. *Asia Pacific Viewpoint* 43, 43–62.

Teppo, A. and Millstein, M. (2015) The place of gentrification in Cape Town. In Lees, L., Shin, H. B. and López-Morales, E. (eds.) *Global gentrifications: Uneven development and displacement.* Policy Press, Bristol, pp. 419–40.

Tomba, L. (2004) Creating an urban middle class: Social engineering in Beijing. *The China Journal* 51, 1–26.

Tsai, M-C. (2001) Dependency, the state and class in the neoliberal transition of Taiwan. *Third World Quarterly* 22(3), 359–79.

UN Habitat (2013) *Streets as public spaces and drivers of urban prosperity.* UN Habitat, Nairobi.

___ (2003) *The challenge of slums – Global report on human settlements 2003.* UN Habitat, Nairobi.

Urry, J. (2007) *Mobilities.* Polity, Cambridge.

Vainer, C., Bienenstein, R., Tanaka, G., De Oliveira, F. and Lobino, C. (2013) O plano popular da Vila Autódromo, uma experiência de planejamento conflitual. *Anais, Encontros Nacionais da ANPUR* 15.

Velásquez Atehortúa, J. (2014) Barrio women's invited and invented spaces against urban elitisation in Chacao, Venezuela. *Antipode* 46(3), 835–56.

Vicario, L. and Monje, P. (2003) Another 'Guggenheim effect'? The generation of a potentially gentrifiable neighbourhood in Bilbao. *Urban Studies* 40(12), 2383–400.

Vigdor, J. (2002) Does gentrification harm the poor? In Gale, W. G. and Pack, J. R. (eds.) *Brookings-Wharton Papers on urban affairs,* pp. 133–73.

Visser, G. and Kotze, N. (2008) The state and new-build gentrification in Central Cape Town, South Africa. *Urban Studies* 45(12), 2565–93.

Wacquant, L. (2007) Territorial stigmatization in the age of advanced marginality. *Thesis Eleven* 91(1), 66–77.

Wacquant, L., Slater, T. and Pereira, V. B. (2014) Territorial stigmatization in action. *Environment and Planning A* 46, 1270–80.

Wai, A. W. T. (2006) Place promotion and iconography in Shanghai's Xintiandi. *Habitat International* 30(2), 245–60.

Walker, D. (2008) Gentrification moves to the global south: An analysis of the Programa de Rescate, a neoliberal urban policy in Mexico City's Centro Histórico unpublished PhD thesis, University of Kentucky.

Walker, R. (2015) Building a better theory of the urban: A response to 'Towards a new epistemology of the urban?' *City* 19(2–3), 183–91.

Wang, J. and Lau, S. (2009) Gentrification and Shanghai's new middle-class: Another reflection on the cultural consumption thesis. *Cities* 26(2), 57–66.

Wang, S. W-H. (2011) The evolution of housing renewal in Shanghai, 1990–2010: A 'socially conscious' entrepreneurial city? *International Journal of Housing Policy* 11(1), 51–69.

Wang, Y. P., Wang, Y. and Wu, J. (2009) Urbanization and informal development in China: Urban villages in Shenzhen. *International Journal of Urban and Regional Research* 33(4), 957–73.

Ward, K. (2009) Toward a relational comparative approach to the study of cities. *Progress in Human Geography* 34(4), 471–87.

____ (2008) Capital and class. In Hall, T., Hubbard, P. and Short, J. R. (eds.) *The Sage Companion to the City*. Sage Publications, Los Angeles, pp. 109–22.

Ward, P. (1993) The Latin American inner city: Differences of degree or of kind? *Environment and Planning A* 25(8), 1131–60.

Warde, A. (1991) Gentrification as consumption: Issues of class and gender. *Environment and Planning D* 9, 223–32.

Watson, V. (2014) African urban fantasies: Dreams or nightmares? *Environment and Urbanization* 26(1), 215–31.

Watt, P. (2009) Housing stock transfers, regeneration and state-led gentrification in London. *Urban Policy and Research* 27(3), 229–42.

Weber, R. (2002) Extracting value from the city: Neoliberalism and urban redevelopment. *Antipode* 34(3), 519–40.

Wilson, W. J. (1987) *The truly disadvantaged: The inner city, the underclass, and public policy*. University of Chicago Press, Chicago.

Winkler, T. (2009) Prolonging the global age of gentrification: Johannesburg regeneration policies. *Planning Theory*, 8(4), 362–81.

Woo-Cumings, M. (1999) (ed.) *The developmental state*. Cornell University Press, New York.

Wright, M. (2014) Gentrification, assassination and forgetting in Mexico: A feminist Marxist tale. *Gender, Place and Culture* 21(1), 1–16.

Wu, F. (2004) Residential location under market orientated development: The process and outcomes in urban China. *Geoforum* 35, 453–70.

___ (2002) Sociospatial differentiation in urban China: Evidence from Shanghai's real estate markets. *Environment and Planning A* 34(9), 1591–615.

___ (2000) The global and local dimensions of place-making: Remaking Shanghai as a world city. *Urban Studies* 37(8), 1359–77.

Wu, F. and Zhang, J. (2007) Planning the competitive city-region: The emergence of strategic development plan in China. *Urban Affairs Review* 42(5), 714–40.

Wyly, E. and D. Hammel (1999) Islands of decay in seas of renewal: Housing policy and the resurgence of gentrification, *Housing Policy Debate* 10, 4: 711–71.

Wyly, E., Newman, K., Schafran, A. and Lee, E. (2010) Displacing New York. *Environment and Planning A* 42, 2602–23.

Yang, J. (2010) The crisis of masculinity: Class, gender, and kindly power in post-Mao China. *American Ethnologist* 37(3), 550–62.

Yang, Y-R. and Chang, C-h. (2007) An urban regeneration regime in China: A case study of urban redevelopment in Shanghai's Taipingqiao area. *Urban Studies* 44(9), 1809–826.

Yeoh, B. S. A. and Huang, S. (1996) The conservation-redevelopment dilemma in Singapore: The case of the Kampong Glam historic district. *Cities* 13(6), 411–22.

Yeung, H. W-c. (2009) Regional development and the competitive dynamics of global production networks: An East Asian perspective. *Regional Studies* 43(3), 325–51.

___ (2000) State intervention and neoliberalism in the globalising world economy: Lessons from Singapore's regionalisation programme. *The Pacific Review* 13(1), 133–62.

Yip, N-m. and Jiang, Y. (2011) Homeowners united: The attempt to create lateral networks of homeowners' associations in urban China. *Journal of Contemporary China* 20, 735–50.

Zhang, J. (1997) Informal construction in Beijing's old neighbourhoods. *Cities* 14(2), 85–94.

Zhang, J. and Peck, J. (2014) Variegated capitalism, Chinese style: Regional models, multi-scalar constructions. *Regional Studies* DOI: 10.1080/00343404.2013.856514

Zhang, L. (2010) *In search of paradise: Middle-class living in a Chinese metropolis.* Cornell University Press, Ithaca, New York.

Zhang, Y. (2013) *The fragmented politics of urban preservation: Beijing, Chicago, and Paris*. University of Minnesota Press, Minneapolis.

Zhang, Y. and Weismann, G. (2006) Public housing's Cinderella: Policy dynamics of HOPE VI in the mid-1990s. In Bennett, L., Smith, J. L. and Wright, P. A. (eds.) *Where are poor people to live? Transforming public housing communities.* M. E. Sharpe, New York, pp. 41–67.

Zukin, S. (2010) *Naked city: The death and life of authentic urban places.* Oxford University Press, New York; Oxford.

Index

Page numbers in *italic* denote an illustration

Abu Dhabi 171
  Guggenheim 122
  mega-projects 38–9, *39*
accumulation by dispossession
  18–19, 189, 195
accumulation by ground rent
  dispossession 169–70
Africa 22, 83, 95, 140, 211, 215
Alexandri, G. 591
Amar, P. 22
anti-gentrification activism/resistance
  20, 164, 218–22
  against evictions 163–7
  and rent gap 74–5
Anti-Gentrification Handbook for
  Council Estates in
  London 45
appropriation 76, 185, 218
  imperialist 208–9
Arabindoo, P. 144
Argentina 76
Ascensao, E. 142
Asian Coalition for Housing Rights
  183–4
Asian financial crisis 57, 192
Athens 59–60
Atkinson, R. 4
Aylesbury and Heygate estates
  (UK) 197

'back to the city' movement 24
Bahrain 171
Ballard, R. 16
Bangalore (India)
  IT corridor installation 41
Barcelona
  22@Barcelona Project 120–1
  Barcelona Model 116–21
Bartolome, L. J. 176
Bavisker, A. 89–90
Beauregard, R. 8
Beijing (China)
  historic preservation 107
  mega-gentrification and
    displacement 172
  Olympic Games (2008) 172, 175
  rapid city-wide redevelopment 154
Beirut 37–8, 39–40
Betancur, J. 115, 138, 204, 221
bid-rent theory 25, 26
Bilbao Model 116, 121–2
Blanco, I. 75–6
Blaut, J. 4
blighting 70
blockbusting 65, 69
Boddy, M. 47–8
Booyens, I. 137
Bose, P. 91
bourgeois environmentalism 90

INDEX   257

Bratton, W. 125
Brazil 66, 160
   Growth Acceleration programme (PAC) 158
   land value capture policy 224
   middle class 85
   see also Rio de Janeiro
Brenner, N. 114
BRIC countries 84–5
Bridge, G. 4, 87
Bristol 48
Britain *see* UK
broken windows thesis 125
brownfield sites 46
Buck-Morss, S. 226
Buenos Aires (Argentina) 68, 98, 225
   mutual-aid housing cooperatives 225
   Puerto Madero redevelopment 119–20, *119*
   slums 141, 144–5
Butler, T. 4, 24, 47, 86, 111

Cabannes, Y. 166–7
Cairo (Egypt)
   peripheral shanty towns 145
   suburban urbanization 43–4
Calcutta 91, 217
Cape Town 113, 161–2
capital accumulation 11, 33, 37, 43, 46, 51, 57, 58, 80, 86, 178
capital reinvestment 30, 31
capital switching 57, 58, 59, 60, 77, 80–1, 142–3, 174–5, 179–80
capitalism 6, 12, 17, 56–7, 210
   as creative destruction 56, 58
capitalized ground rent (CGR) 61–2
Cartier, C. 185
centrality 29, 77

centre-periphery relationship 7, 8, 27, 50, 77
chain displacement 190, 192, 195
Cheonggye Stream Restoration Project (Seoul) 178–9, *179*
Chicago
   historic preservation 107
   mixed communities policy 130
Chicago School 7, 11, 25, 25–6
Chile 20, 61, 66
   campamentos 59
   capital switching 598
   household structures 98
   middle classes 97–8
   neoliberalization 58–9, 115
   poblaciones 144
   rent gap 64
   resistance to evictions in Pedro Aguirre Cerda (PAC) 164–5
   see also Santiago
China 19, 66, 171, 204, 209, 212–13
   capital switching 60
   creative city policy 135, 135–7
   displacement caused by Cultural Revolution 213
   displacement and metro system construction in 176
   displacement scale 191
   gender relations 96
   historic preservation and gentrification 103–6, 107
   'land-based accumulation' 186
   mega-gentrification and displacement 172–3, 185–6, 199
   middle classes 85, 86, 88, 89, 97, 103, 213, 220
   redevelopment of former rural villages 72–3

China (cont.)
    speculative urbanization 16
    see also individual cities
Choi, N. 77, 176
city (urban) peripheries
Ciudad Juárez (Mexico) 63
Clark, E. 29, 31, 74, 168
Clarke, N. 16–17
class 83–8 *see also* middle class
Clinton, Bill 17
Cochrane, A. 116
COHRE 190
Colombia 98, 222
    land value capture policy 224
commercial gentrification 178
community land trusts 223
comparative urbanism/urbanists 2, 3, 6, 12–13, 17, 202, 206, 207
complete urbanization of society 76
concentric zone model 11, 25
consumption 95–100, 109
Contreras, Y. 97–8
council estates 45, 196–7, 214, 224
    redevelopment/state-led gentrification of in UK 196–7, *198*
counter-urbanization 45–6
creative city 114, 133–7
creative destruction 54, 56, 70
    and capitalism 58, 210
creative gentrification 112
critical political economy approach 6, 13, 15
critical urban learning 222
crowd politics 89
'culture of property' 19, 37, 95
Cunningham, S. 135, 137

Darling, Eliza 46
Datta, A. 42–3
Davidson, M. 48–9, 87, 191, 217
Davis, L.K. 184
Davis, Mike 22, 141
De Beukelaer, C. 37
de-distancing 12
debt crisis 56
decommodification of housing 222
defamation of place 70–1
deindustrialization 46, 58
Delhi (India)
    Bhagidari scheme 90–1
    remaking of through judicial orders 89–90
dependency 6, 22
Desai, V. 142–3, 168–9
development studies 7
    and urban studies 7
development-led displacement (DID) 174, 175–6, 177, 179
    infrastructure development 177–8
    major characteristics of 176
developmental state 3, 6, 53, 115
Dharavi Redevelopment Project (DRP) 151, 152
Dharavi Slum (Mumbai) 148, *148*, 150
diffusionist logic 112–13
displacement 9, 17, 41, 45, 48–9, 50, 61, 139, 146, 213–14
    chain 190, 192, 195
    development-led *see* development-led displacement (DID)
    exclusionary 190–1, 195
    hidden (global South) 190–4
    importance of previous histories 213–14
    last-resident 190, 191
    mega- 172, 174–6, 195–7, 198, 209, 214

INDEX 259

phenomenological 9, 49, 197
and mega-gentrification 171–200
and re-urbanization 48–9
and redevelopment in
 Sanggyedong 183–5
and rent gap 54, 61
rural villages 41–2
and slum gentrification in Rio de
 Janeiro 156–8
see also evictions
displacement pressure 191
dispossession 41, 42
divided city, concept of 146, 147
Dongguanli project (China) 187–9
dos Santos Junior, O. 156
Doshi, S. 22, 149, 152, 199
Dubai 17, 38, 171
Durban 141
Dutton, P. 113

East Asia 20–1, 37, 53, 79, 81, 115,
 200
and the creative city 135–6
gay gentrification 99
housing sector 180–1
mega-gentrification and
 developmental states 180–5
middle class 85–6, 92
see also individual countries
eco-city 114
economic crisis (2008) 81, 140
economy
 gentrification and global 56
Ecuador 102
 Malecón 2000 waterfront
  regeneration
  (Guayaquil) 119, 120
Eko Atlantic megaproject (Lagos,
 Nigeria) 17
elites 15

Elshahed, M. 43, 44
emancipatory city thesis 95
endogenous gentrification 205
Engels, F. 109
 The housing question 140
entrepreneurial city-building 43
entrepreneurial state 79
Euro-American gentrification 20, 86,
 206, 207
Europe 19–20
evictions
 forced 44, 157, 166, 175, 190, 198,
  199
 resistance against 163–7
 Sanggyedong 183–5
exclusionary displacement 190–1,
  195

Fangzhicheng Integrated
 Development Office 187
Fangzhicheng project 187–9
favela 32, 76, 144, 155–9, 169, 209
Fernandes, L. 91, 99
financial crisis (2008) 81, 140
financialization 46, 146–7, 195, 212
Floor Area Ratios (FARs) 66
Florida, Richard 133–4
Foshan (China)
 redevelopment in 187–9

gated communities 22, 26, 40
gay gentrification 98, 98–9
gender 96
generalized gentrification, rise of 36
gentrification 30
 alternatives to 222–6
 blueprints 113, 217
 characteristics of 28–9
 coining of phrase by Ruth Glass
  1, 10

gentrification (cont.)
  in the context of planetary
    urbanization 31–40
  definition 9, 28, 30–1, 201
  economics 54
  enabling and necessary factors 204
  focus on inner-city areas 30–1, 49
  and the global economy 56–60
  and redevelopment 28–9, 30
  relational understanding of
  resistance to 18, 114, 218–22
  urban development characteristics
    influencing 30
gentrification frontier 10
'gentrification generalized' 4, 5
'gentrification of the mind' 94
gentrification models 116–22
  Barcelona Model 116–18
  Bilbao Model 116, 121–2
gentrification policies 113–14,
    122–37
  creative city policy 114, 133–7
  mixed communities policy 128–33
  zero-tolerance policing 124–7, 132
gentrifier 10, 24, 26, 44–5, 47, 51,
    66, 74, 76, 83–110, 134, 145,
    152, 169, 180, 204, 211–13,
    224
geographical switching 34
Germany 20
Ghertner, A. 90
GINI indices 146
Giuliani, Rudolph 125–6, 127
Glass, Ruth 1–2, 3, 7, 8, 10, 21, 25,
    29, 51, 173, 201
  'Gaps in Knowledge' 2, 10
  'London: Aspects of Change' 1
global class 108
global economy
  and gentrification 56–60

global elites 86
global gentrifier 84, 87, 109, 212
global gentrification
  blueprint 111–39
global habitat 87
global suburbanisms 40–6
global urbanism 12
globalization 207
  of urban policy 116
Goertz, Geoffrey 84
Goetz, E. G. 196
Goldman, M. 32, 41, 42
González, S. 117, 118
governance 15–16, 42, 90, 107,
    116–17, 123, 136, 145,
    149–50, 165, 180, 209,
    225
Greece 59–60, 94
Greenwich Village (New York
    City) 100
Grimsrud, G. M. 45
ground rent 60, 64, 69, 75, 81–2,
    163, 165, 169 209–10, 224
Guayaquil (Ecuador) 102
  Malecón 2000 waterfront
    regeneration 119, 120
Guggenheim Effect 122
Guggenheim Museum 122

Hackworth, J. 66, 218
Harris, A. 147, 149
Hart, G. 13
Harvey, David 18, 33, 57, 139,
    174–5, 189
Haugaard, M. 199
Haussmannization 50, 197 *see also*
    neo-Haussmannization
hegemony, of dominant interests 199
Heidegger, M. 12
heritage conservation 102, 108

hidden displacement 190–4 *better under displacement*
historic preservation 100–8
Hong Kong 19, 36–7, 71–2, 72, 171, 205
  middle class 85
  property boom 110
  public housing provision 37
  and transit-oriented development (TOD) 177–8
  Urban Renewal Agency 71–2
  vertical accumulation 38
HOPE VI programme (US) 129–31, 195–6
housing 78
  decommodification of 222
  social/public 37, 59, 66, 79, 80, 129, 131–2, 143, 151, 181, 195, 214, 221, 223
  unaffordability 69
Housing and Urban Development (HUD) (US) 129, 130, 132
Huang, L. 79
hutongs 103–4
hybrid gentrification 80, 110, 162, 169
Hyra, Derek 214

Ibero-American Centre for Strategic Urban Development (CIDEU) 118
IMF (International Monetary Fund) 115
inclusive city 146
India 32
  and Barcelona Model 121
  mega-urbanization in cities 43, 43
  middle class 90, 91
  mixed communities policy in Kolkata 128–9

slums 147, 148–9
see also Delhi; Mumbai
informal districts/settlements 66, 72, 75–7, 132, 141–2, 148, 152–3, 155, 160, 174–5, 177, 179, 183, 214–15,
  gentrification of formerly 160–3
informal housing 144, 215 *see also* slums
informality 3, 115, 151, 156, 205, 214
infrastructure development
  and gentrification 177–9
injustice 207
inner-city areas 24–7, 28, 49
  focus on 49
  regeneration of 46–7, 49
  and rent gap 61–2
Inter-American Development Bank (IDB) 16, 101, 115
Inzulza-Contardo, J. 97, 138
Istanbul (Turkey) 101
  slums 141, 145
  Taksim Square anti-gentrification protests 219, 220

Jacobs, Jane 100
Johannesburg (South Africa) 22, 113, 132
joint redevelopment programme (JRP) (Seoul) 67–8, 181–3, 192, 198–9

Karachi 141, 145, 221
Karnaaka Industrial Areas Development Board (KIADB) 42
Keil, Roger 11, 40, 51–2
Kharas, Homi 84
Kim, S.-H. 21, 31, 36, 67, 182
Kingsley, G. T. 196

Kolkata (India)
  mixed communities policy 128–9
Kong, L. 135
Koo, H. 91, 92–4
Kotze, N. 112–13
Krijnen, M. 37

laissez faire 59, 60
Lama-Rewal, S.T. 91
Land Acquisition Act (1894) 42
land-based accumulation 186
land grabbingland rent 60–2
  monopoly of land value capture policies 224–5
Larner, W. 116
Latin America 20, 45, 54
  and Barcelona Model 118–20
  gentrification in 138
  historic preservation and gentrification 101–2
  land value capture 54
  middle classes 83–4, 85, 96
  neoliberalism 115
  resistance to gentrification 221–2
  sexual politics 98–9
  see also individual countries
Lau, S. 97
Lebanon 38
Lees, L.
  new build gentrification 48, 217
  displacement 49, 191
  spatial capital 76–7, 160
  emancipatory city thesis 95
  new middle class in China 103
  gentrification of council estates 197
  reconceptualization of gentrification 206
Lees, Shin and Lopez-Morales
  Global Gentrifications: uneven development and displacement
Lees, Slater and Wyly
  Gentrification 31
  mutation of gentrification 8
  Ruth Glass 10
  Chicago School of urbanism 25
  rural gentrification 33
  The Gentrification Reader
Lefebvre, H. 2, 10, 33, 34, 35, 58, 226
  The Urban Revolution 33
Lemanski, C. 91, 162, 169
lesbian gentrification 98
Lett, D. 94
Ley, David 3, 19, 36, 37, 86, 94, 95, 205
lifestyle 95–100
Lim, H. 178
Lima 222
'liveable' city agendas 89–90
localized action plans 223, 224
Loftus, A. 142–3, 168–9
London 1, 20, 45
  Aylesbury Estate *131*
  coining of term 'gentrification'
  Docklands regeneration 47
  inflated property market 110
  redevelopment of council estates and displacement 196–7, *198*
  slums 215–16, *216*
  super-rich 110
López-Morales, E.
  displacement 45, 64–65, 77
  floor area ratio 66
  gentrification by ground-rent dispossession 32, 169, 209
  inner-city redevelopment 32, 144, 164

INDEX

redlining 66, 209
rent gap 64–6
resistance to gentrification 164–165
Los Angeles 142

Ma, L. 2, 21
McCann, E. J. 116, 118, 139
McFarlane, Colin 5, 12, 13, 226
Mahigeer Tahreek (Pakistan) 221
Malécon 2000 waterfront redevelopment (Guayaquil, Ecuador) 119, 120
Manila (Philippines) 77
   rail project and displacement 175, 177
Maragall, Pasqual 117
Marcuse, Peter 48, 49, 190, 222
marginality
marginal gentrification 152
Marx, Karl 57, 60, 109
Marxism/Marxists 13, 14–15, 18, 94
Massey, D. 6
Mbembe, A. 22
mega-displacement 172, 174–6, 195–7, 198, 209
   and public housing estates 195
   and slum clearance 214
mega-events 41, 154, 160 *see also* Olympic Games
mega-gentrification 31
   China 172–3, 185–9, 199
   and displacement 171–200
   and East Asian developmental states 180–5
   and hidden displacement in the global South 190–4
   in neo-authoritarian socialist states 185–6
mega-projects 17, 29, 38–9

Mehta, Mukesh 150, 151
Merrifield, Andy 8, 10, 19, 27, 35, 50, 77, 88, 89
Mexico 208
   neoliberalism 115
Mexico City 16
   Programa de Rescate (rescue programme) 126–7
   Sante Fe area 160–1
   zero-tolerance policing and 'Plan Giuliani' 125–7
middle class(es) 74, 212
   East Asia 85–6
   emergence of 83–4
   emergence and rise of 'new' 84–5, 109, 212
   as global gentrifiers 83–8, 212
   and politics 89–95
   and the state 88
Middle East 22, 37–8
   see also individual countries
Millstein, M. 113
mixed communities policy 128–33, 143, 217
mobility 75–6
Monclús, F.J. 118
Monje, P. 121
Montevideo (Uruguay) 225
Morocco, slums 141
Mueller, G. 78, 79
Mumbai (India) 17, 22, 142, 145
   Dharavi Slum 148, *148*, 150, 151, 152
   mega-urbanization 42, 43
   slum gentrification 147–52, *148*, 199
mutual-aid housing cooperatives 225
Myers, Gareth 202, 211
Mysore-Bangalore Information Corridor (India) 32

Nangok (Seoul) neighbourhood redevelopment 192–4, 193, 194
Nathan, R. 26
neighbourhood transformation 51
neo-authoritarian socialist states 185
    mega-gentrification 185–6
neo-Haussmannization 35, 43, 50, 56
neoliberal urbanism/urbanization 7, 35, 128
neoliberalism/neoliberalization 53, 60, 114–16, 123, 145, 163, 217
    Chile 58–9, 115
    and the creative city 135
    Latin America 115
new build gentrification 21, 42, 47, 67, 68, 108, 110, 173, 217
New Deal for Communities regeneration programme (UK) 196
New Deal (US) 57
new economics 53–82
new urban renewal 214
new urbanism 36, 46–7
New York City 16, 208
    historic preservation 106
    zero tolerance model 124, 125, 127
Nigeria 17
Nogueira, Colonel Celso 127
Norway 45–6
Nuttall, S. 22

Obrador, Andrés Manuel López 125, 126
obsolescence 70, 73
Ocaranza, M. 164

Occupy movements 88
Ocean Estate (Tower Hamlets) 196–7
O'Connor, J. 135–6
OCT-LOFT 136–7
Olympic Games 17
    Barcelona (1992) 116–17
    Beijing (2008) 172, 175
    Rio de Janeiro (2016) 156–7
    Seoul (1988) 181, 184
Organization of American States (OAS) 101

Pakistan 208
    redevelopment projects 135
    resistance to gentrification 221
    slums 141
Panama 64, 102–3
    redevelopment of *Casco Antiguo* 64, 65, 102
Paris 19–20
    Haussmannization 50, 197
    historic preservation 107
Park, B-G. 181
Parnell, S. 12
Peck, J. 16, 112, 115, 116, 134, 138
People's Action Party (Singapore) 89
peripheral urban development 40–6
peripheralization 208, 211 see also centre-peripheral relationship
Pedro Aguirre Cerda (PAC) (Chile) 164–5
planetarity 207
planetary urbanization 10–11, 18–19, 77, 207, 216, 225–6
    gentrification in the context of 31–40
police pacification 158
policies, gentrification see gentrification policies

policy discourse 123
middle-class politics 88–95
Portugal, slums 142
post-crisis gentrification 77–80
postcolonial theory 21
postcolonial urbanism 6–7, 8
potential ground rent (PGR) 62, 210
Poulantzas, N. 94
poverty 141–2
Pradilla, E. 160
primary circuit of capital (industrial production) 16, 19, 33–4, 36, 54, 57, 60, 175
Prince, R. 137
privatization 115, 116
property rights 182, 188, 191
property-based interests 64, 67, 198, 200
property-led redevelopment 121, 154, 186
public housing estates 66, 80
and mega-displacement 195–6
Puerto Rico 128
 Las Gladiolas housing estate 131–2, *133*, 221
 mixed communities policy 130–1
 zero-tolerance approach to crime 132

Qianmen (China) 104, 188
Quito Letter 101–2

re-urbanization 47–9
real estate 15, 42–3, 50, 51, 54, 223
 role of in capitalist societies 33–6
redevelopment 11, 21, 22, 36, 48, 67, 70–3, 75, 80, 102–3, 106, 113, 118–22, 127, 132
 and gentrification 28–9, 30, 31, 38, 40, 51, 55, 59, 172–3, 177, 181–99, 203–22
 and rent gap 32, 61, 63, 64–66, 81
 and resistance 37
 of slums 140–168
regulation 6, 16, 60, 66, 71, 79, 123, 182, 191, 205
relational comparative approach 13–14, *14*
Ren, X. 97, 104–5, 106, 213
rent 54, 60–2, 81
rent gap 32, 44, 61–2, 78, 81, 177, 197
 and anti-gentrification activists 74–5
 Ciudad Juárez 63
 and class struggle 74
 and displacement 54, 61
 and inner city 61–2
 role of state 66
 Santiago (Chile) 32, 45, 59, 64–5, 76, 97–8, 130, 145, 164–5, 209
 Seoul 31, 66–9, 127, 172–3, 178, 181, 183, 192, 198
 and territorial stigmatization 70
 towards a comparative urbanism of planetary 62–70
rent gap theory 54–5, 61–3, 82, 210
 resurgence of 73–5
Rerat, P. 76, 77, 160
residential rehabilitation 28
resistance to gentrification 164, 218–22
 evictions 163–7 eviction is not resistance, put on its own
revanchism 16, 80, 102, 116, 124–5, 208
reverse-gentrification 150

Ribeiro, Queiroz 32, 155–6
Right to the City 163–4, 168, 220, 224–5
right to the urban 225
Rio de Janeiro (Brazil) 16, 31–2, 41, 204, 209
  'favela chic' phenomenon 159
  Metrocables 76
  Olympic Games host (2016) 156–7
    slum/favelas gentrification 150–1, 155–9
    Special Group for Urban Control 208
    Vila Autódromo 41, 157–8
    zero tolerance policing 127
    Zona Portuária 158, 159
Robinson, Jennifer 7, 9, 12, 13, 112, 206
Rodan, G. 89
Rofe, M.M. 86
Rogers, Sir Richard 117
Rojas, E. 118
Rose, Damaris 152, 213
Roy, Ananya 6, 8, 10, 12, 22, 139, 206, 211
Ruiz-Tagle, J. 130
rural gentrification 7, 33, 45
rural villages 72–3, 137, 153,
  land dispossession and displacement 41–2
Russia 79

Samara, T. 145–6
Sandroni, P. 66
Sanggyedong (South Korea), redevelopment 183–5
sanitization of space 16
Santiago (Chile) 32, 45, 59, 67, 97, 98, 145, 209

Efecto Metro 76–7
mixed communities policy 130
rent gap 64–5
resistance to eviction in PAC district 164–5
Sassen, S. 108
Schill, M. 26
Schindler, S. 15–16, 17, 18
Schulman, S. 201
Schumpeter, J. A. 56, 57
securitization 16
secondary circuit of capital (built environment) 4, 16, 19, 33–5, 42, 51, 54, 57, 58
SecoviRio 158–9
Seoul (South Korea) 64, *68*, 173
  Cheonggye Stream Restoration Project 178–9, *179*
  Joint Redevelopment Programme (JRP) 67–8, 181–3, 192, 198–9
  Nangok neighbourhood redevelopment 192–4, *193*, *194*
  new-build gentrification 67, 68–9
  role of state in rent gap formation 66
Serbia
  Belgrade Waterfront 122
shacks
Shanghai (China) 17, 104, 188
  historic preservation 104, 104–5
  mega-gentrification and displacement 172
  middle class 213
  Xintiandi redevelopment 106, 188
Shanghai World Expo (2010) 121
Shaw, W. 133
Shenzhen (China)
  creative city policy 136–7

Shin, H. B.
  speculative urbanization in China 16
  speculative urbanization in South Korea 21, 36
  mega events 17
  metropolitan-scale gentrification 31
  rent gap 66
  new-build gentrification in Seoul 67–8
  mega-displacement 182
  Nangok 192
  cross-class alliance 225–6
Sidaway, J. 4
Sigler, T. 64
Simone, AbdouMaliq 12
Singapore 89
  creative city policy 135
  public housing provision 181
Singerman, D. 22
Slater, T. 54, 74
Slim, Carlos 125
slum gentrification 140–70, 197, 199, 208, 214–17
  China 152–5
  Mumbai 147–52, 148, 199
  Rio de Janeiro 150–1, 155–9
  and the state 74, 168, 170, 215
slum redevelopment and gentrification 152–5
Slum Rehabilitation Scheme (SRS) 151
slums
  definition and characteristics 144
  hidden economic potential of 169
  numbers and location of 141, 145
  use of term 143–4
smart city 114

Smith, Neil 4, 12, 29–30, 36, 61, 111, 124, 138, 171–2, 203, 207, 209, 218
  Gentrification and uneven development 28
  The new urban frontier 30
social cleansing 114
social mobility 130, 212
Solidere 40
Soto, Hernando de 169
South Africa 22, 112–13, 162
  Abahlali movement 218
  mixed income housing 132–3
  slums 140
South Asia 21–2
South Korea 21, 36, 66, 192, 209
  mega-gentrification and displacement 172–3, 209
  middle class 85, 88
  middle classes and politics 91–4, 93
  public housing provision 181
  resistance to gentrification 218–19
  student protests 92–3, 93
  see also Seoul
Soweto (South Africa)
  mixed income housing 132
space
  regulation and securitization of 16
  sanitization of 16
spatial capital 55, 75–7, 160
spatial elitism 4
speculative urbanization/urbanism 6, 21, 36, 37, 50, 66, 67, 95, 189, 211
speculative urban accumulation
Spivak, G. 12, 207
state 88, 100, 204, 223–4
  and historic preservation 100, 104

state (cont.)
    intervention in gentrification 180–9
    key actor in planetary gentrification 109
    state-led gentrification and middle classes 88–91
    role of 15–16
    and slum gentrification 74, 168, 170, 215
state developmentalism (see developmental state)
stigma 81, 131, 144, 156, 195–6, 210
    and devaluation 70–3
Straw, Jack 125
street protests 93–4
sub-prime mortgage crisis 57, 77
Subaltern Studies School 14–15
substandard settlements 66, 181–3, 197–9
suburban gentrification 7, 10
suburban sprawl 46
suburbanization 11, 24, 25, 26–7, 40–6, 61, 95, 96, 205, 211
super-rich 110
Swanson, K. 102
Sweden 224
Sykora, L. 63

Taipei 79, 80
    Da-an Forest Park construction 177
Taiwan 79–80, 85
tenure (restructuring of) 133, 142, 149, 152, 163, 182, 192–3, 196, 209
Teo, S.Y. 3, 19, 36, 37, 205
Teppo, A. 113
Theodore, N. 16, 114, 116, 138
Tiger economies 55

touristification 16
transferable development rights (TDRs) 151
transit-oriented development (TOD) 55, 177–8
transnational elites 40, 53, 199, 204
transnational gentrification 64
transport 25, 55, 57, 75–7, 81, 158, 160–1, 204, 205
Tuan, Yi Fu 49
Turkey 141
    Taksim Square anti-gentrification protests (Istanbul) 219, 220

UAE (United Arab Emirates) 38, 122
UNDP 141
    Human Development Report (2014) 141–2
uneven development 45, 171
    see also rent gap
United Kingdom (UK)
    New Deal for Communities Regeneration programme 196
    redevelopment of council estates and displacement 196–7, 198
    urban renaissance policy 117
    zero-tolerance policing 125
    see also London
United Nations
    Habitat report 140, 144, 168
United States
    critical economy 78
    HOPE VI programme 129–31, 195–6
    see also individual cities
upper-income families 26
urban accumulation 70, 179, 199, 212

urban development 7, 11, 26, 27, 30, 31, 43–6, 50, 66, 96, 108, 118, 129, 149, 171, 186, 199, 213, 216, 222
  characteristics influencing gentrification 30
  peripheral 40–6
urban entrepreneurialism 55
urban policy 7, 29, 37, 116, 134
  globalization of 116, 139
  making 29, 37, 113
  transfer 111–12, 202
urban regeneration models 116–22
Urban Renewal Agency (Hong Kong) 71–2
urban renewal schemes 17–18
Urban Revolution 58
urban rights 18
  right to the city
urban–rural differences 10
urban–rural breakdown of binary
urban social movements 18, 51
urban studies
  'comparative urbanism' in 2, 3, 6, 12–13, 17, 202, 206, 207
  'cosmopolitan turn' in 12
  and development studies 7
Urban Task Force (UK)
  Towards an Urban Renaissance 117
urbanism
  Chicago School of Sociology 25
urbanization 2, 205
  counter- 45–6
  dominance of Euro-American discourse 2–3, 8
  planetary *see* planetary urbanization
  speculative 16, 21, 36, 37, 50, 66, 67, 95, 189, 211
  spectacular 38

Urry, J. 75
use value 165, 171

Vainer, C. 156
Vancouver (Canada) 89, 90, 136
Vicario, L. 121
Vila Autódromo project (Rio de Janeiro) 41, 157–8
Visser, G. 112–13
vulture international funds 78

Wang, J. 97
Ward, K. 116
Ward, P. 20
Waschsmuth, D. 64
Washington Consensus 138
Washington DC 11, 16, 78–9
Watt, Paul 196
Weber, Rachel 58, 70
weiquan 219
wilderness gentrification 46
Wilson, W. J. 130
Winkler, T. 113
World Bank 16, 115
world cities 50
worlding 12, 50
Wright, M. 63
Wu, F. 2, 21
Wyly, Elvin 190

Xi'an (China), redevelopment of 187, 188–9
Xintiandi (China) 106, 188

zero tolerance policing 102, 124–7, 132, 210, 217
Zhang, Y. 106–7
Zimbabwe 140
Zukin, Sharon 5, 9, 117